Decision-Making and Analytics

In today's data-driven business world, organizations struggle to extract actionable insights from vast amounts of information. This textbook bridges theory and practice, offering structured approaches to leverage data for enhanced decision-making through machine learning, AI, and optimization modeling – transforming raw data into competitive business intelligence.

Decision-Making and Analytics: Data-Driven Business Intelligence and Optimization features advanced analytical methodologies paired with real-world implementation studies, creating a progressive learning path from customer behavior analysis through hybrid decision models. By integrating fuzzy logic, neural networks, text mining, and reinforcement learning, it provides a comprehensive toolkit applicable across business scenarios, turning abstract concepts into practical solutions.

Serving multiple audiences, this resource bridges the gap between classroom theory and industry practice for students in industrial engineering, business analytics, and management information systems. Practitioners gain applicable methodologies to improve organizational decision-making. At the same time, researchers benefit from cutting-edge approaches across various domains, and policymakers can develop data-informed strategies, addressing the growing demand for professionals who translate data into strategic decisions.

This textbook employs open-ended questions that develop analytical reasoning skills without predetermined solutions, better preparing students for real-world complexity. Supporting materials include high-quality figure slides, as well as PowerPoint slides for qualified adopters, enabling engaging learning environments that effectively communicate complex concepts while building practical, data-driven decision-making skills.

Decision-Making and Analytics

Data-Driven Business Intelligence and Optimization

Hamed Fazlollahtabar

CRC Press
Taylor & Francis Group
Boca Raton London New York

CRC Press is an imprint of the
Taylor & Francis Group, an **informa** business

Designed cover image: Shutterstock

First edition published 2027
by CRC Press
2385 NW Executive Center Drive, Suite 320, Boca Raton FL 33431

and by CRC Press
4 Park Square, Milton Park, Abingdon, Oxon, OX14 4RN

CRC Press is an imprint of Taylor & Francis Group, LLC

ISBN: 978-1-041-23733-4 (hbk)
ISBN: 978-1-041-24116-4 (pbk)
ISBN: 978-1-003-74014-8 (ebk)

DOI: 10.1201/9781003740148

Typeset in Times
by Newgen Publishing UK

Access the Instructor and Student Resources: www.routledge.com/9781041237334

Contents

Preface

In today's rapidly evolving business landscape, organizations face increasing complexity in decision-making due to vast amounts of data and dynamic market conditions. The concept of big data, artificial intelligence, and machine learning provided a new paradigm in systems optimization. Data analytics and data-driven optimization facilitate real-time decision-making. *Decision-Making and Analytics: Data-Driven Business Intelligence and Optimization* provides cutting-edge methodologies and analytical approaches to enhance industrial and business intelligence, optimize operations, and drive data-informed strategies. The book is suitable as a textbook for decision-making and analytics courses in different disciplines, including industrial engineering and management.

As a textbook, this book explores innovative techniques in industrial and business analytics, including machine learning, fuzzy logic, social network analysis, and optimization models, to address real-world challenges. It begins with advanced customer behavior analysis using regression-based fuzzy expert systems (Chapter 1) and extends to social network analytics for e-commerce insights (Chapter 2). Subsequent chapters delve into AI-driven customer segmentation (Chapter 3), churn prediction via integrated neural networks (Chapter 4), and text mining for customer relationship management (Chapter 5).

The book also covers intelligent recommender systems (Chapter 6), multi-factor analysis for product development (Chapter 7), and cloud-based ERP solutions for SMEs (Chapter 8). Further, it examines social network applications in insurance analytics (Chapter 9), IoT-enhanced loss functions for telecom networks (Chapter 10), and reinforcement learning for energy management (Chapter 11). Finally, it presents hybrid decision-making models like the Analytic Network Process and Best-Worst Method for cybersecurity in e-commerce (Chapter 12).

This textbook encompasses a wide range of implementation studies in all chapters to illustrate the application areas of each proposed methodology for decision-making and analytics.

Designed for undergraduate and graduate students, practitioners, researchers, and policy makers, this book bridges theory and practice, offering actionable insights for leveraging data-driven intelligence in industrial and business optimization. Whether addressing customer engagement, operational efficiency, or strategic innovation, it equips readers with the tools to navigate modern business complexities with confidence.

About the Author

Hamed Fazlollahtabar is Associate Professor in the Department of Industrial Engineering at Damghan University, Iran, where he has served since June 2017. He earned his BSc and MSc in Industrial Engineering from Mazandaran University of Science and Technology in 2008 and 2010, followed by a PhD in Industrial and Systems Engineering from Iran University of Science and Technology, and he completed a postdoctoral research fellowship in reliability engineering for complex systems at Sharif University of Technology from October 2016 to March 2017. A globally recognized researcher, Dr. Fazlollahtabar has been consistently ranked among the top 2% of scientists worldwide from 2021 through 2025, was named the best researcher across all engineering disciplines in Iran in 2022, and received the Distinguished Young Researcher Award in Industrial Engineering in Iran for 2023. His extensive research portfolio encompasses robotic production systems, reliability engineering, sustainable supply chain planning, and business intelligence and analytics, resulting in over 300 research papers published in international books, journals, and conferences, as well as 12 authored books, 8 of which have international academic distribution. Dr. Fazlollahtabar further contributes to his field through service on editorial boards of academic journals and participation in technical committees for conferences.

1 Regression-Based Fuzzy Expert System for Customer Behavior Analysis

1.1 INTRODUCTION

With its increasing growth, the Internet has found many applications in different industries. The Internet and related technologies have created new and different types of trade and business, and the time spent on the Internet, social networks, and online shopping is increasing rapidly. Today, the Internet has become a powerful media for the advertising industry. And in a society where every business tries to have its own space for competition, every business uses different online strategies to target their customers. The best business strategy attracts more new customers, even customers of competitors. Due to the many features and advantages that the Internet has compared to traditional media, online advertising has witnessed a lot of growth in recent years. Internet advertising involves a publisher who integrates the ads into online content and an advertiser who provides the ad display. Display advertising can be seen in various forms such as text, images, video, and audio. In an Internet advertisement, an advertiser pays a publisher a fee to allocate space on their website to display their banner or advertisement and attract the attention of users who visit that page. Companies have increased their desire to increase brand awareness and advertising and make their products known. In fact, users are more likely to buy products from brands they trust and know. Displaying ads creates an emotional experience that delights the user and increases their trust. The emergence and spread of social networks have created new markets in the electronic world, and managers have long sought to understand how advertising affects consumer decisions.

A business starts with the consumer and their needs, and an organization can survive only when it can meet the needs and desires of the consumer with a correct and comprehensive understanding. Investigating the impact of Internet advertising, especially through social networks, on consumer behavior can provide information to business owners that is effective in improving customer interaction.

The purpose of this research is to design a mechanism to evaluate customer behavior under the influence of advertising through the Telegram messenger. The type of customer behavior in the market has always been of interest to economists, manufacturers, and sellers. How does a customer decide to choose a product or service of their choice? What are the parameters that mainly affect a customer's purchase? It is generally considered that economists mainly develop a mathematical model for

DOI: 10.1201/9781003740148-1

customer behavior based on their logical thinking and adjust the development of a production strategy based on this model. Sometimes the customer's emotions play a stronger role than their logical thinking in decision-making. In many cases, contrary to the imagination of entrepreneurs and business developers, customer behavior for buying is not coordinated and consistent with the framework of the product suppliers' imagination. This means that in designing a new business and evaluating the market for a product, we always look at a range of customers who have different tastes and purchasing budgets, and our idea is that we can cover all tastes and budgets by offering a range of products. Consumer behavior is the study of how people buy, what they buy, and why they buy in the first place. Consumer behavior is a sub-branch of marketing that consists of elements such as psychology, marketing, and economics. In fact, we study whether a customer will buy in response to advertising or not. Now we want to do this structure with the help of an expert system. The computational tool in our expert system is regression.

Investigating the impact of online advertising, especially through social networks, on consumer behavior can provide information to business owners that is effective in improving customer interaction. Given the popularity it has gained, the Telegram social network can become a suitable platform for improving business on the Internet.

The following assumptions are considered in this study:

- The number of users is certain, and the behavioral assessment interval is predetermined;
- The users are characterized by having an education related to computers and information technology;
- Only commercial advertisements are considered;
- In evaluating and collecting behavioral information of users, only the first reaction is considered, and no reversal of the decision is allowed;
- Learning from past behavior is allowed.

1.2 SOCIAL NETWORKS AND ADVERTISING

Social networks keep users connected through sharing and building friendships, and the high intimacy between users is a good place for word-of-mouth marketing. A social network is a platform where users of groups can become friends with each other and have casual interactions. Social networks developed from traditional offline networks to online social networks due to the Internet revolution. This migration from offline networks to online networks can inherit many of the features of offline networks. Social attraction is the most important factor that is similar between the two, and even many online friendships eventually extend to offline relationships (Fazlollahtabar and Ehsani, 2010).

Marketers and customers today extend their communication through the new dynamic medium called social networks. And this is the latest development in product advertising and customer communication. Facebook in particular is one of the social networks that is increasing in users every day, which includes numerous brainstorming sessions that are spontaneous among its users to develop a theory. This powerful social media platform is a prime opportunity for any brand to promote their product through exposure, attention, and perception to develop their brand and create value (Fazlollahtabara and Mahdavi, 2012).

Many studies that focus on social media users' attitudes toward advertising have reported that consumers have a strong tendency to avoid advertising. Investigating this behavior is important for both advertisers and websites because they are influenced by consumer attitudes. Avoidance of advertising occurs in three ways: for example, when reading a magazine or newspaper, we intentionally skip the advertising section (cognitive method), or when a commercial is broadcast on television, we leave the room (behavioral method), or we close the advertising banners on the computer (mechanical method).

Younger users are more likely to trust online sources than older users because they live in an age where technology has always been there. Before modern technologies emerged, it was possible to avoid advertising by ignoring or redirecting it. But now, with the advancement of technology and computers, it has become easier to hide advertising from view.

A virtual social network is a chain of connections and nodes of a social network or the non-physical presence of individuals in a virtual location. In a virtual social network, communication between individuals and organizations or groups is not face-to-face and without intermediaries, but depends on a tool called the Internet. These days, when we talk about social networking on the Internet, we are actually talking about a virtual and intangible network of human communication. In a relatively short space of time, the use of social networks and group sites has found a permanent place in the daily lives of many people. In recent decades, the availability of user-generated documents on social media has increased dramatically. This huge growth of user-generated content has affected traditional shopping. Social networks enable customers to communicate directly with sellers and then make a purchase decision. Discovering the tastes and feelings of users regarding different products and brands, and identifying the habits of buyers for buying a particular product, is traditionally a costly and time-consuming task. However, new techniques have been provided for extensive market research. Also, using these techniques and analyzing the content created by the user, as well as the user's activity and behavior in the user profile, can be very effective in analyzing and achieving the goal. For example, a seller can track the user's behavior from the time they select a product and in monitor the final purchase of the desired product (Farjoo et al., 2018).

Today, we live in the era of communication and the expansion of communication web applications – the most influential of which are Facebook and Google+. These programs allow individuals to introduce themselves individually and communicate with others. The competition between these social networks is intensifying, and the main issue for these social networks is the replacement and reduction of users. Since the income of these social networks is through advertising, increasing or decreasing users is very important for them. In order to fully use advertising to attract customers, business people must know the customer's attitude toward text messages, emails, and social networks so that they can plan advertisements, such as content and advertising timing, etc. This customer behavior can be identified by analyzing customer response to multimedia advertising. Telegram is one of the social networks whose number of users is increasing every day. Marketing on social networks ultimately leads to electronic word-of-mouth advertising, which makes the value and credibility of your marketing message much greater than in other cases, and newer customers trust you're advertising much more easily, because when one of their acquaintances shares your marketing message, the person receives your advertising message through one of his friends and acquaintances, and this means that your message has been approved by an acquaintance (Fazlollahtabar, 2021a).

Studies on consumer attitudes toward advertising media have been conducted in several countries, but no research has been conducted on the comparison of the main antecedents of different advertising media. Of all the factors, acceptability and credibility are more influential than others in consumer attitudes and behavior toward advertising via social media.

1.3 RESEARCH METHODOLOGY

In this chapter, the paradigm used to conduct the research "Analysis of Online Customer Behavior in the Influence of Telegram Advertising" is first explained, and its four pillars are described in the following order: ontology, epistemology, methodology, and value theory. Here, with regard to the area under study, classification, summarization, and extraction of relationships between concepts were discussed. The research methodology is explained in terms of research strategies, purpose, type, time, society, statistical sample and sampling method, data collection method, stages of completing the research, method of obtaining the components and concepts of the conceptual model of the research, stages of analyzing online customer behavior in the influence of Telegram advertising and its structure, tools for determining the components of the decision-making model, and how it is valid and reliable (Fazlollahtabar, 2021b).

In fact, the issues of this research can be explained as the ambiguity and fatigue of Tehran online store managers in analyzing online customer behavior in the influence of Telegram advertising; And also, the issue of the need for multiple expertise through the simultaneous application of the knowledge of experts in different fields to solve the problems of analyzing online customer behavior in the effectiveness of Telegram advertising was expressed. In the present study, analyzing online customer behavior in the effectiveness of Telegram advertising using a regression-based expert system, under the title T-OCBA.REG-ES, is presented for the first time in a research field related to the subject (Fazlollahtabar and Ehsani, 2010).

After library studies, content classification, and gap analysis, we will collect materials and review the research background. Then, the indicators for examining Telegram user behavior are determined with the help of marketing factors. The design of the expert system and its information flow mechanism are designed and the related information system is drawn under the business intelligence architecture. Multivariate linear regression is used to make decisions. These decisions are made by examining the behavior of users during the data collection period. Expert systems refer to a special category of computer software that strives to assist human experts and specialists or partially replace them in limited specialized fields. Such systems are, in fact, the initial and simpler examples of more advanced technology of knowledge-based systems. These systems usually store information in the form of facts and rules in a structured form in a database called a knowledge base, and then, by using special expert-inference methods, the required results are obtained from this data. Expert systems are programs that mimic the behavior of a human expert in a particular field. This program uses the information that the user stores in them to express an opinion on a particular topic. Hence, expert systems continue to ask you questions until they can find a topic that matches your answers. In statistical models, regression analysis is a statistical process for estimating relationships between variables. This method includes many techniques for modeling and analyzing specific and unique variables, focusing on the relationship between the dependent variable and one or more independent variables. Regression analysis is particularly helpful in understanding how the value of the dependent variable changes when any of the independent variables change, while holding other independent variables constant. The most common application of regression analysis is to estimate the conditional mean of the dependent variable from a set of independent variables, which is equivalent to the mean of the dependent variable when the independent variables are fixed. The least common application is to focus on the quantile or spatial parameter of the conditional distribution of the dependent variable from a set of independent variables. In all cases, the goal is to estimate a function of the independent variables, called the regression function. In regression analysis, the dispersion of the dependent variable around the regression function is determined, which can be described by a probability distribution. Regression analysis is widely used for forecasting. Regression analysis is also used to understand the relationship between the independent and dependent variables and the shape of these relationships. In certain situations, this analysis can be used to expertly infer excellent relationships between the independent and dependent variables (Fazlollahtabar and Ehsani, 2010). However, this can lead to false or invalid relationships, so caution is advised. Familiar methods such as linear regression and least squares are parametric, in that the regression function is estimated from the data under a limited number of unknown parameters. Nonparametric regression refers to methods that allow regression functions to be fitted to a given set of functions with an infinite number of parameter probabilities. A general linear model is a statistical linear model. It can be written as follows:

$$Y = XB + U \tag{1.1}$$

where Y is a matrix with rows of multivariate measurements, X is a matrix that can be a design matrix, B is a matrix containing the variables that are usually estimated, and U is a matrix containing errors and noise. The errors are usually assumed to be uncorrelated with the measurements and to follow a multivariate normal distribution. If the errors do not follow a multivariate normal distribution, a generalized linear model is used to make the assumptions about Y and U more reasonable.

1.3.1 Research Design

In fact, in the present study, datasets and the behavior of information systems users and interviews with experts are used. The study population of this research can be categorized into two general groups: the first group includes expert professors, and the second group includes managers and IT and industrial engineering experts of Tehran online stores.

In fact, the sampling method in this research is a combination of two methods: non-probability purposeful (judgmental) sampling and snowball sampling. In this sampling method, the researcher can, in selecting the members of the sample population, preferentially select samples that are in a more favorable situation in terms of contextual components and the level of familiarity with the area under study. In the end, it should be said that the sample size of this research is people who are available and willing to cooperate, in total, from the two groups introduced in the statistical population.

1.3.2 Research Model

In general, the applied algorithm is called Mamdani or Min-Maximum, which is abbreviated below. The rules of the Mamdani inference method are explained later. Suppose the fuzzy input value $\tilde{x} = A'$ then:

1. Calculate the degree of realization of the desired rule from Equation (1.2);

$$B_i = max_x \left[\mu_{A'}(x)^\wedge \mu_{Ai}(x)\right] \tag{1.2}$$

2. Calculate the output of the fuzzy set B'_i for each rule using t-norm:

$$B'_i : \mu_{B'_i}(y) = B_i{}^\wedge \mu_{B_i}(y), y \in Y, 1 \le i \le k \tag{1.3}$$

3. Merge the output of the fuzzy sets B'_i

$$B'_i =: \mu_{B'_i}(y) = max_{1 \le i \le K} \mu_{B'_i}(y), y \in Y \tag{1.4}$$

Note, if the inputs of A′i are also fuzzy sets, then the degree of realization is calculated using Equation (1.5).

$$\begin{aligned} B'_i = {} & max_{x_1 \epsilon X_1}\left[\mu_{A'_1}(x_1)^\wedge \mu_{A_{i1}}(x_1)\right]{}^\wedge max_{x_2 \epsilon X_2}\left[\mu_{A'_2}(x_2)^\wedge \mu_{A_{i2}}(x_2)\right] \\ & {}^\wedge \ldots {}^\wedge max_{x_p \epsilon X_1}\left[\mu_{A'_p}(x_p)^\wedge \mu_{A_{ip}}(x_p)\right] \end{aligned} \tag{1.5}$$

Generic steps:

Step 1: Determining appropriate criteria

Two types of criteria are considered for the credit rating of customers in organizations based on their field of activity. One: Quantitative criteria that are determined using definitive data available to the organization. Two: Qualitative criteria that are determined from expert opinions and using verbal variables. In this study, all numbers will be fuzzy and triangular in type because they are based on direct understanding and transfer and have an effective computational representation.

Then, Chang's method will be used to determine the priority level. In the following relations, G is the list of criteria from number 1 to n, and M represents the priority value of the criteria over each other; Mj_{gi} (j = 1, 2, …, M), and all values used in the model are fuzzy.

$$G = \{g_1, g_2, g_3, \ldots, g_n\}; M^1_{gi}, M^2_{gi}, \ldots, M^m_{gi}\ i = 1, 2, \ldots, n.$$

The steps of Chang's solution for calculating the fuzzy weight are as follows:

In the first step, using the fuzzy sum operator, we obtain the sum of each row of the pairwise comparison matrix (i.e., for each criterion) based on Equation (1.6).

$$\sum_{j=1}^{m} M_{gi}^{j} = \left(\sum_{j=1}^{m} I_j, \sum_{j=1}^{m} M_j, \sum_{j=1}^{m} U_j \right) \tag{1.6}$$

In the second step, we add the sum of the fuzzy numbers obtained from the first step together.

$$\sum_{i=1}^{n} \sum_{j=1}^{m} M_{gi}^{j} = \left(\sum_{j=1}^{m} I_j, \sum_{j=1}^{m} M_j, \sum_{j=1}^{m} U_j \right) \tag{1.7}$$

In the third step, we calculate the inverse value of the vector obtained in the second step.

$$\left[\sum_{i=1}^{n} \sum_{j=1}^{m} M_{gi}^{j} \right]^{-1} = \left(\frac{1}{\sum_{j=1}^{m} I_j}, \frac{1}{\sum_{j=1}^{m} M_j}, \frac{1}{\sum_{j=1}^{m} U_j} \right) \tag{1.8}$$

The fourth step is to multiply the fuzzy value obtained by summing the rows of the first step by the value obtained in the third step.

$$S_i = \sum_{j=1}^{m} M_{gi}^{j} \otimes \left[\sum_{i=1}^{n} \sum_{j=1}^{m} M_{gi}^{j} \right]^{-1} \tag{1.9}$$

In this way, the fuzzy weight vector of each criterion will be obtained.

Step 2: Determining the structure of the fuzzy expert system

In order to fuzzify the inputs, specific ranges are considered for the qualitative criteria of linguistic values and for the quantitative criteria, and their membership functions are defined. Then, to form the knowledge base, knowledge engineering is carried out. By conducting in-depth interviews and numerous brainstorming sessions, the knowledge and experiences of experts are collected, and the rules of the expert system are created. The next step is to determine the type of inference method of the fuzzy expert system. The inference center of an expert system is formed by sets of if-then rules. An optimal reasoning method is to first infer each rule and then integrate the results, which is called FITA. Among all the FITA reasoning methods, the Mamdani inference method is very widely used and common, which has also been used in this research.

The fuzzy expert system model of this research is of the "multiple input – single output" (MISO) type, which can be converted into a set of rules in the form of relation (1.10):

$$\begin{aligned} &R_i = IF\ X_1\ is\ A_{i1}\ AND\ IF\ X_2\ is\ A_{i2}\ AND \ldots X_j\ is\ A_{ij} \\ &THEN\ y\ is\ B_i, \mathrm{i} = 1, \ldots, \mathrm{n} \end{aligned} \tag{1.10}$$

So that in this relation;

R_i is the i-th rule, n is the number of rules, X_j (j = 1, …, r) is the number of input variables, y is the only output variable of the system, Aij is the input of fuzzy sets defined in the reference set space U = U1, …, Ur, and Bi is the output of fuzzy sets defined in the reference set V.

Therefore, each rule is a local fuzzy relation between U × V that maps a part of the multidimensional input space U to a specific part of the output space V. As shown in Equation (1.10), the relationship between all inputs of this study is in the form of AND. The integration of the rules of Equation (1.10) forms the basis of the system rules, which are presented in the form of Equation (1.11):

$$R = \bigcup_{i=1}^{n} R_i = R_1 ALSO\ R_2\ ALSO \ldots ALSO\ R_n \tag{1.11}$$

In this study, the t-norm and t-conorm operators have been selected for the AND and ALSO connections, respectively, so that from the family of operators mentioned, Dr. Zadeh operators have been used. The values of the obtained outputs are in the form of fuzzy numbers, which are defuzzified by the phase center of gravity method to simplify their analysis.

Step 3: Determining the value of the criteria for evaluating online customer behavior

The first stage of processing a fuzzy expert system is the fuzzification of system inputs. To prepare the inputs and then fuzzify the input variables, the following steps must be performed. Quantitative criteria can be calculated and are obtained as definite data from reports and data available to the organization. However, qualitative criteria are quantified by experts with verbal variables. To obtain a unit value for each qualitative criterion, the verbal variables must be converted into triangular fuzzy numbers so that they can be calculated and analyzed. Therefore, a decision matrix must be formed for the qualitative criteria so that a unit value can be obtained for each qualitative criterion. We calculate the average of the experts' opinions using (1.12) and place the final value for each quality criterion in the decision matrix.

$$AVG_1 = \frac{\sum_{i=1}^{n} R_{1i}}{n} \tag{1.12}$$

1.4 EXPERT SYSTEM METHODOLOGY BASED ON REGRESSION

One of the most important reasons for using fuzzy regression logic in this research is that real-world problems usually have a complex structure, which indicates the existence of ambiguity and uncertainty in their definition and understanding. Since humans were able to think, they have always faced ambiguity in various social, technical, and economic problems. Even the invention of the computer and its practical development in analyzing real-world problems could not solve the problem of ambiguity and uncertainty. Despite inaccurate and insufficient information, humans can approximately understand and analyze the behavior and nature of the system when facing complex real-world problems. The human brain defines and evaluates sentences by considering various factors. Based on expert-inference thinking, modeling them in mathematical language and formulas would be a very complicated, if not impossible, task. Linguistic components are expressed based on linguistic (discourse) values, which are in the set of expressions (words/terms), and linguistic expressions are attributes for linguistic components. Here, linguistic components are referred to as components for which the accepted values are words and sentences of human and machine languages instead of numbers. A fuzzy regression number is a particular fuzzy regression set of the following form, where x accepts real values of the members of the set R, and its membership function is as follows:

$$A' = \{(x, \mu,(x)) | x \in X\} \tag{1.13}$$

The regression-based expert system for analyzing online customer behavior in the influence of Telegram advertising (T-OCBA.REG-ES) has the following components:

a. Regression fuzzifier: A regression fuzzifier is located at the input of the system that converts the numerical value of the components into a fuzzy regression set. The reason for not using bell-shaped and nonlinear membership functions in this research is that the most commonly used fuzzy regression numbers are triangular and trapezoidal fuzzy regression numbers. Triangular fuzzy regression numbers are defined by three real numbers expressed as (l, m, u), where the value m is the most probable value of a fuzzy regression number. They have a membership function that is constructed based on sparse information, and it is easy to perform four operations on these numbers, which are often used in cases such as fuzzy regression controllers, managerial, commercial, and financial decision-making, comparisons, and evaluations. A triangular fuzzy regression number A with a linear piecewise membership function μ_A is defined as the following relationship:
b. $\mu_A(X) = 0$ if $x<a1$; $\mu_A(X) = x\text{-}a1/a2\text{-}a1$ if $a1 \leq x \leq a2$; $\mu_A(X) = a3\text{-}x/a3\text{-}a2$ if $a2 \leq x \leq a3$; $\mu_A(X) = 0$ if $x > a3$. A graphical presentation for a triangular fuzzy number is shown in Figure 1.1.
c. Fuzzy Regression Rule Base: A fuzzy regression rule base is a set of if-then rules that is considered the heart of a fuzzy regression system because other

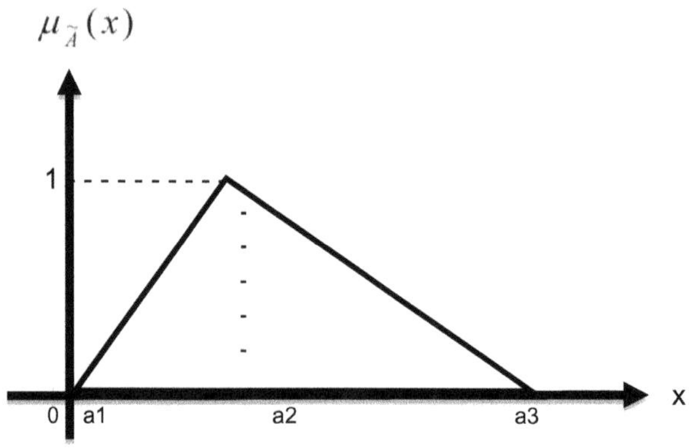

FIGURE 1.1 A triangular fuzzy number.

components of the fuzzy regression system are used to implement these rules effectively and efficiently.

d. Fuzzy Regression Inference-Expert Engine: The fuzzy regression inference-expert engine converts inputs into outputs with a series of operations and uses some of the following fuzzy regression operators:
Join (And) (T-norm): If B, A are two arbitrary sets, then their join is equal to the minimum of the membership function of the set B, A, and it is represented as A∩B:

$$\mu_{A\cap B}(x) = \min(\mu_A(x), \mu_B(x)) \text{ and } T(x,y) = x.\ y \tag{1.14}$$

Join (Or) (S-norm): If B and A are two arbitrary sets. The union B, A is equal to the maximum membership function of the set B, A and is represented as $\mu_{AUB}(x)$:

$$\mu_{AUB}(x) = \max(\mu_A(x), \mu_B(x)) \text{ and } S(x,y) = x + y - x.\ y \tag{1.15}$$

Complement (Not) (C-norm): If S is a set and A is a subset of it. The complement of the set A is the result of subtracting all members of A from one and is represented by Ā or $\mu_{\text{not }A}(x)$

$$\mu_{\bar{A}}(x) = 1 - \mu_A(x) \tag{1.16}$$

e. Non-regression fuzzifier (vari-regression fuzzifier): The fuzzifier transforms the output of the fuzzy regression into a definite number. The task of the non-regression fuzzifier is to determine the point that best represents the fuzzy regression set B'. The formula for the central mean is as follows:

$$\text{DeFuzzifier} = \Sigma^m_{l=1}\, y^{-1}\, W_{1/}\, \Sigma^m_{l=1}\, W_1 \tag{1.17}$$

In this relation, y is the center of the fuzzy regression set and w_L is its height degree. In this way, for the considered relations, the weights between the relations between the concepts can be obtained.

Considering the application of the expert system using fuzzy regression logic calculations designed in this research, five stages for analyzing online customer behavior in the influence of Telegram advertising using the regression-based expert system (T-OCBA.REG-ES) was considered as follows:

1. Modeling the concepts of the field of online customer behavior analysis in the influence of Telegram advertising in order to identify the input and output components and draw the relationships between them.
2. Defining qualitative components using linguistic adverbs and assigning fuzzy regression numbers and sets and membership functions to them.
3. Presenting a fuzzy regression model based on the basis of the definitions and designs made using MATLAB software: This stage includes extracting the rules of expertise and evaluating them by experts and creating a fuzzy regression rule base as well as designing an expert-inference engine with access to the fuzzy regression rules.
4. Designing the user interface and how to display the options, and how to use the designed regression-based expert system.
5. Choosing a method for defuzzification regression to convert the numbers and sets of fuzzy regression into a definite value to actually examine the system's performance.

1.4.1 Expert System Validation

In the present study, to collect information, datasets and behaviors of information system users, data collection tool No. 1 (tool for measuring the impact of effective variables on analyzing online customer behavior in the effectiveness of Telegram advertising using a regression-based expert system) and data collection tool No. 2 (tool for validating a regression-based expert system for analyzing online customer behavior in the effectiveness of Telegram advertising), and interviews are used. Before using the measurement tools, the researcher must be sure of the validity of the measurement tool in question. In order to examine the extent to which the measurement tool of this research measures the desired characteristic, scientific documents, and standards for analyzing online customer behavior in the effectiveness of Telegram advertising in Tehran online stores, as well as the opinions of supervisors and consultants and some experts of Tehran online stores, have been used, and the tool for determining the components of the initial decision-making model has reached its final form after applying their opinions. The reason for examining the reliability of the data collection tool of the aforementioned research is to determine the extent to which the measurement tool produces the same results under the same conditions. Therefore, Cronbach's alpha is used here to calculate the reliability coefficient of the measurement tool. In order to validate the fuzzy regression expert system for analyzing online customer behavior in the effectiveness of Telegram advertising in Tehran online stores, the outputs and responses of the "T-OCBA.REG-ES system" were compared with the average opinions of experts in the field under study using a separate tool.

On the other hand, the human brain defines and values sentences by considering various factors and based on inferential-expert thinking, which would be a very complicated, if not impossible, task to model in language and mathematical formulas. Linguistic variables are expressed based on linguistic (discourse) values, which are in the set of expressions (words/terms), and linguistic expressions are attributes for linguistic variables. Here, linguistic variables are variables whose acceptable values are words and sentences of human and machine languages instead of numbers. The programming space "MATLAB" is a software environment for performing numerical calculations and a fourth-generation programming language. The word MATLAB means both a numerical computing environment and the corresponding programming language itself, which is created from the combination of the two words matrix and laboratory. This name indicates the matrix-oriented approach of the program, in which even single numbers are considered as matrices. In fact, the advantage of using fuzzy regression logic in this research is that it connects the space of inputs to the space of outputs with a mapping, and the basic principles for doing this are a series of if-then statements that we call rules. All rules are evaluated in parallel, and the order of the rules is not important at all. The rules themselves are important because they all refer to variables and describe variables. The fuzzy regression systems section of MATLAB software consists of the following sections:

Input: In this section, the system inputs are specified.
Inference Engine: In this section, the rules are written.
Output: In this section, the system outputs are determined, which, for our work, is "Analysis of online customer behavior in the effectiveness of Telegram advertising using a regression-based expert system".

1.5 IMPLEMENTATION STUDY

According to the model presented in the previous section and the relevant implementation steps, an implementation study is reported in this section. What is important is the importance of the impact of social networks on customer behavior. Therefore, the environment under study is necessarily the online shopping market in Iran.

The present study is carried out with the help of 120 people who have had experience using Telegram advertising. For this purpose, first the required variables that were collected before are surveyed using fuzzy linguistic variables. These opinions are called independent variables in multivariate linear regression. To evaluate the output of the survey results, the fuzzy rules analysis environment of MATLAB software is used.

To identify the criteria, first the subject literature and internal and external sources related to customer validation were reviewed. The criteria presented in the subject literature were related to online customers, so after collecting the criteria from different sources, nine variables were identified. To eliminate ambiguity and ensure uniformity in the perception of each criterion group, a precise definition was specified for them in the aforementioned meeting. In this study, the expert system was

TABLE 1.1
Linguistic Variables Related to the Analysis of Online Customer Behavior

Number	Linguistic Variable	Membership Function Abbreviation	Triangular and Trapezoidal Numbers
1	Very low	V-L	(0, 0, 0.03, 0.15)
2	Low	L	(0.1, 0.2, 0.3)
3	Medium	M	(0.3, 0.5, 0.7)
4	High	H	(0.7, 0.8, 0.9)
5	Very high	V-H	(0.8, 0.85, 0.97, 1, 1)

designed in five stages, which are: fuzzification, fuzzy rule base, fuzzy inference engine, defuzzification, and validation of the designed expert system.

Implementation Steps:

Fuzzification: After designing the conceptual model of the research, the input and output variables of the expert system are defined. In fact, the module "Analysis of Online Customer Behavior in the Influence of Telegram Advertising" forms the output of the proposed expert system. Table 1.1 shows the distribution of linguistic variables related to the output variable of the research expert system based on the experts' opinions.

Fuzzy rule base: The fuzzy rule base is a set of if-then rules that is the heart of the expert system of this research. Here, the probability of occurrence of different states between the main variables of the expert system is considered to be the same. The method of generating the knowledge base rules of the main module of the expert is shown in Figure 1.2.

Calculating the weight of each of the main variables using the experts' opinion: According to the final weight table of the measuring variables, the weight of each of the main variables was obtained, which can be seen in Table 1.2.

Calculating the value of the output variable based on the weight of each variable: Considering the weight of each of the nine main variables of the expert system, the behavior of online customers in being affected by Telegram advertisements can be examined in different modes. Table 1.3 shows how to calculate the weight of possible modes for generating a rule in the expert system knowledge base. Other knowledge base rules of this expert system were also generated in this manner, and a questionnaire was used to compare the expert system's responses with the average opinion of the experts.

Fuzzy inference engine: In the Fuzzy Box section of MATLAB, the fuzzy rule aggregation style is selected to connect the fuzzy rules of the system using the "and" and "sum" operators. Thus, the more accurate sum of each output set of rules is considered, not their maximum. The main reason for using the Mamdani fuzzy inference system is that the Mamdani fuzzy inference engine has overcome the disadvantages of the pure fuzzy system and the Sugeno fuzzy system, while placing a fuzzifier at the input and a non-fuzzifier at the output of the system. Prod is used to

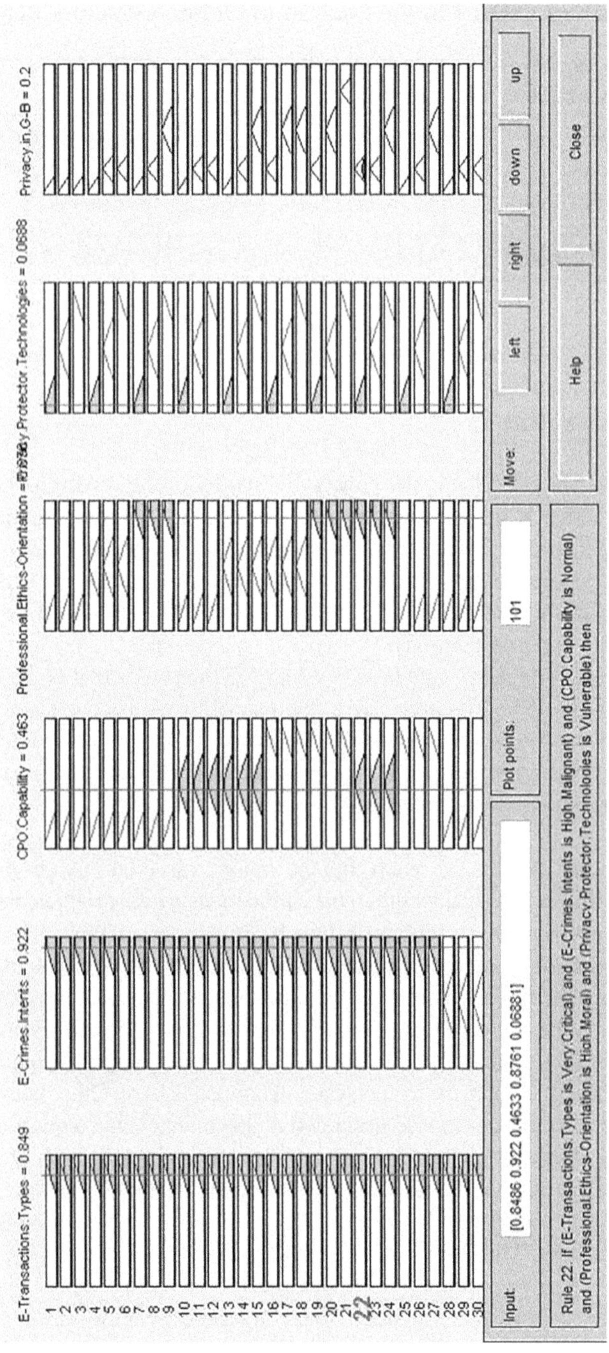

FIGURE 1.2 Fuzzy rule base toolbox.

TABLE 1.2
Weight of Each of the Main Variables of the Study

Number	Main Variables	Measures	Final Weight
1	Access	7	0.611
2	Gender	18	0.479
3	Age group	14	0.472
4	How often does one visit Telegram	11	0.471
5	How much time does one spend on Telegram each time	11	0.568
6	Educational level	16	0.436
7	Monthly income	19	0.667
8	Marital status	10	0.432
9	How many times has one made online purchases	17	0.765

TABLE 1.3
How to Calculate the Weight of Possible States for Generating a Rule in the Expert System Knowledge Base

Possible States for Generating a Rule	Effect on the Output Variable	Weight of the Variable × Definite Value of the Variable	Weight of the Assumed State
If "accessibility" is very high	The negative relationship	s 0.611 × 0.025	0.0152
and "gender" is in the medium position	The positive relationship	0.479 × 0.5	0.2395
and "age group" is high	The positive relationship	0.472 × 0.975	0.4602
and "how often does he come to Telegram" is very high	The negative relationship	0.471 × 0.025	0.0117
and "how much time does he spend on Telegram each time" is low	The positive relationship	0.568 × 0.025	× 0.0142
and "education level" is medium	Positive relationship	0.436 × 0.02	0087/0
and "monthly income" is very high	Positive relationship	0.667 × 0.03	0200/0
and "marital status" is low	Negative relationship	0.432 × 0.25	0108/0
and "how many times have you made online purchases" are very low	Negative relationship	0.765 × 0.025	00191/0
Then	Average weight of the assumed states		
What is the state of online customer behavior in terms of being affected by Telegram advertising?		0.0152 + 0.2395 + 0.4602 + 0.0117 + 0.0142 + 0.0087 + 0.0200 + 0.108 + 0.0191 = 0.8966 ÷ 9 = 0.0996	

select the type of implication in MATLAB, because the minimum operator makes the output fuzzy set short and incomplete. In fact, the reason for using Prod is to more accurately scale the output of fuzzy sets. The defuzzifier converts the fuzzy output into a definite number. In the defuzzifier section of the MATLAB software, the centroid method is selected.

Testing the model is done by entering the R_L and R_S values for all the significant risk factors in a project into the fuzzy expert system; for each risk factor, the R_M values are calculated by the fuzzy expert system, and by summing all the R_M values for all the risk factors, the contingency cost is obtained.

Then the contingency values obtained from the system are compared with the actual contingency values for the completed projects, and the model error is calculated. The reason for choosing this project was the access to its information through the literature and access to several project experts. The statistical population of this study is 120 experts in this field.

Conceptualization: This process is carried out by combining the values of R_L and R_S using the AND operator. In this case, the AND operator is used to connect the preceding parts. Two rules are extracted from the fuzzification process described above:

1. if R_L is M and R_S is M, then R_M is M
2. if R_L is M-L and R_S is M, then R_M is M-L

Since the AND operator is used as the minimum membership value as the degree of membership, in this case, the combination between R_L = M(0.2) and R_S = M(1) yields the value of R_M = M(0.2). Similarly, for R_L = M-L(0.8) and R_S = M(1), the value of R_M = M-L(0.8) is obtained. The aggregation process is also performed to combine the results of the previous stage. The result of the aggregation is in the same area as the previous step.

Defuzzification: It is necessary to defuzzify the result of the aggregation process, which is a fuzzy set. For this model, the (COA) method is selected as the defuzzification method, which is calculated from Equation (1.18).

$$Z_{COA} = \frac{(3.5*0.8)+(4*0.8)+(4.5*0.8)}{(0.8+0.8+0.8)} = 4 \tag{1.18}$$

R_M values for other risk factors can also be obtained using a similar method. To obtain the M_R value of each major risk group, the R_M results obtained for all risk factors in each group are put into Equation (1.1), then the resulting M_R values are put into the regression equation, and the impact of the project variables is obtained.

$$\begin{aligned} cc^\circ = {} & 1.2346 + 0.38 * x_1 + 0.23 * x_2 + 0.32 * x_3 + 0.49 * x_4 + 0.70 * x_5 \\ & + 0.38 * x_6 + 0.43 * x_7 + 0.51 * x_8 + 0.43 * x_9 \end{aligned}$$

According to the above equation, the importance of each of the effective variables in Telegram advertising has been determined. For a better evaluation, in the next section, we will analyze the results. This is to validate the regression results.

Final Evaluation of Expert System Responses is collected after designing the research expert system by randomly selecting 10 rules from the expert system; the outputs of the expert system were compared in a separate questionnaire with the opinions of 18 experts. As found out from the results, the final difference between the outputs of the expert system and the average of the experts' opinions is not significant (0.04675). This result shows that the fuzzy expert system of Telegram advertising factors has sufficient validity.

1.6 CONCLUSION AND DISCUSSIONS

In this study, we analyzed the behavior of online customers in the effectiveness of Telegram advertising using a regression-based expert system. The purpose of this study was to evaluate the behavior of users toward advertising and its effectiveness. An expert system was used to mechanize decision-making and analysis. This expert system used fuzzy theory to consider uncertainty in response using verbal variables. This study was conducted with the help of 120 experts who had experience using Telegram advertising. For this purpose, the required variables that had been collected were first surveyed using fuzzy linguistic variables. To evaluate the output of the survey results, the MATLAB software fuzzy rule analysis environment was used.

In this research, the expert system was designed in five stages, which are: fuzzification, fuzzy rule base, fuzzy inference engine, defuzzification, and validation of the designed expert system. After completing the above steps and obtaining the regression equation, the designed system was evaluated using a group of 18 experts, and the results were validated. The final difference between the outputs of the expert system and the average of the experts' opinions was not significant (0.04675). This resulted in the conclusion that the fuzzy expert system of Telegram advertising agents has sufficient validity.

Future Suggestions:

To develop the research method proposed in this research and considering the growing trend of social networks, the following topics are suggested for other researchers:

- Developing the learning process in an expert system with the help of a reinforcement learning algorithm;
- Designing an internal messenger based on the outputs of this research in the e-commerce space;
- Designing a decision support system for making real-time decisions in the field of smartening businesses based on advertising on social networks;
- Developing B2C and B2B business models based on the results of this research with the approach of increasing market volume.

2 Customer Online Shopping Analysis Using Social Networks and Levenshtein Distance

2.1 INTRODUCTION

The e-commerce environment, which is based on the Internet, allows customers to search for information and purchase goods and services through direct communication with online stores. It should be noted that online shopping is not based on the actual experience of buying goods, but on appearances such as images, shapes, qualitative information, and advertisements of goods. In this environment, we seek to achieve a suitable return and provide various methods to help achieve this.

The ultimate goal of any business is to be profitable, and for this, better results can be achieved by analyzing behavior and managing customer relationships. Customer behavior refers to the buying habits of an individual, including social trends, frequency patterns, and background factors that influence their decision to buy anything. The study of customer behavior is conducted in any business to understand the target audience and create more attractive products, services, and shopping offers. Customer behavior does not describe who shops in your stores, but rather how they shop in your stores. These factors include what customers buy, what products they prefer, and how they perceive your marketing, sales, and customer service. Knowing these details helps businesses connect with customers in a great and enjoyable way. Customer behavior analysis provides insight into the various variables that influence the audience. It takes into account the customer's motivations, preferences, and decision-making methods. This analysis helps us understand how customers feel about our company and whether this perception aligns with their core values (Fajoo et al., 2018).

Now the question is, how can we identify these feelings to ultimately identify customer buying behavior?

The profitability and market share of any business are based on the number of relevant customers.

Sellers are always trying to attract customers by studying customer buying behaviors and predicting their behavior by creating different strategies. This can help sales and marketing departments to create more meaningful relationships with their customers and thereby achieve more sales, which results in better productivity.

The purpose of this research is to create a model to identify customer buying behaviors in different businesses and, as a result, improve the sales situation, market

DOI: 10.1201/9781003740148-2

share, and profitability of the business. The selected method is based on the PPM framework; and the Levenshtein algorithm, which, by examining the factors affecting customer behavior, gives us an output for building a suitable strategy for identifying customer buying behavior.

The following assumptions are considered in this research:

- Customer buying indicators governing the conventional e-commerce space in the world are collected;
- The required data are of a definite and available type;
- Customer buying behavior is considered rational.

The study of customer behavior includes studying how a customer purchases a product, when they purchase it, why they buy it, where they buy it, and at what intervals and frequency they purchase it. A customer-centered approach based on enlightened principles is necessary and essential in today's diverse global market, as a result of which a company is able to identify and serve its target market, reduce dissatisfaction, and get ahead of its competitors. End consumers often make purchases for personal consumption, within the context of their home and family. A consumer's purchasing behavior is also influenced by social factors, including the groups to which the consumer belongs and is influenced by. Every culture includes a number of subcultures (people with shared values). Subcultures can include nationalities, religions, racial groups, or groups of people who share the same geographic location, all of which influence purchasing behavior.

Customer switching is the phenomenon where customers switch from one provider of a similar service, such as a coffee service, to another. Using customer order records, it is possible to identify customer behavior patterns, and this can also be achieved by focusing on customer behavior changes between the target store and other stores that offer the same service. Change is a part of customer behavior, and a complete understanding of customer behavior can play an important role in creating tactics as well as strategic plans. To successfully carry out customer retention activities, it is important for us to correctly understand customer behavior changes and identify the causes of the problem.

CRM should also focus on predicting customer behavior. Accurately predicting customers' future decisions, which allows companies to take appropriate actions in a timely manner, is a very important issue. As the cost of switching from one online store to another becomes lower (easier), customers may switch between stores routinely, making it difficult to identify the main (important) switching behavior. The costs associated with switching from one product to another are incurred by the customer and may be borne by the customer, the company, or both. These costs are not limited to financial aspects but can also be related to psychological aspects that affect brand loyalty.

Types of switching costs:

- Transaction costs
- Equipment compatibility

- Learning to use new services or equipment
- Uncertainty about the quality of new services and the cost of searching for new suppliers or customers

These issues make it more expensive for companies to attract new customers than to retain existing customers, so customer retention is a more appropriate strategy for generating profits. Regression is one of the most common methods in predicting this issue due to its high accuracy and understanding of customer behavior.

The statistical method of logistic regression estimates the relationship between input and output variables. The values of the dependent variables are usually 0 or 1 and are encoded in regression models. In terms of customer churn, the probability is $y = 1$, otherwise $y = 0$.

Genetic algorithms and decision trees are frequently used to select the most important features that affect customer turnover. A set of customer-related features is analyzed using decision-tree methods. Another method used to identify customer switching behavior is the chi-square method. The chi-square method is used to measure expectations from real data compared to observed data; of course, this data must be random and raw. This method is based on identifying the parameters that affect customer behavior, scoring each factor, and forming a conclusion table. The customer score is calculated based on the customer's profession and the quality of the review. If the customer's profession is related to the product and has additional information about the product, the quality of the review will also be higher. Other methods, including PPM and the Levenshtein algorithm, are available for analyzing customer behavior change.

2.2 BACKGROUND

Fazlollahtabar and Ehsani (2010) pointed out that customer relationship management contributes significantly to long-term profitability. Fazlollahtabar et al. (2010) pointed out that retaining existing customers is much more profitable than attracting new customers because sellers need to spend more money to attract new customers, such as discount offers and trust-building costs, so it is preferable to retain existing customers. In customer relationship management (CRM), it is also necessary to work on predicting customer behavior. Accurately predicting customers' decisions in the future allows companies to take appropriate actions in a timely manner. Switching costs are related to changing from one product to another by the customer, and this cost may be incurred by the customer, the company, or both. These costs are not limited to financial aspects only but can also be related to psychological aspects that affect brand loyalty.

Telecommunication service companies use a wide range of tactics, such as creating a service portfolio, offering diverse programs to meet customer needs, offering discounts to ensure customer loyalty, and attracting new customers as a result of increasing market share. Market share is inversely proportional to the churn rate, i.e., the higher the churn rate, the higher the market share. Factors such as brand strength, marketing efforts, and quality issues may affect market share. Changing is a part of customer behavior, and a complete understanding of customer behavior can play an important role in creating tactics as well as strategic plans.

To successfully carry out customer retention activities, it is important for us to understand customer switching behavior correctly and identify the reasons for the problem. Researchers pointed out that the PPM framework has been widely used in human migration research. Since there are many similarities between migration behavior and customer switching behavior and both can be attributed to the phenomenon of moving from one place to another, it is claimed that this framework is very effective in changing customer behavior (Fazlollahtabar and Mohseninejad, 2015).

Fazlollahtabar (2021b) studied a mobile cloud service provider and examined it in the PPM framework. The findings show that two Push factors, six Pull factors, and two Mooring factors affect the switching behavior of this company's customers. It is also claimed that the PPM method can be effective for better understanding the customer and the factors that influence the customer. The results of these studies show that all the factors of the PPM framework are effective to some extent in understanding customer behavior change.

2.2.1 Summary and Explanation of the Gap

In this research, we try to examine customer purchasing behavior and analyze it using a combination of PPM methods and the Levenshtein distance algorithm, which is a proposed and new method for investigating this issue. In the following sections, we will examine these methods and combine them to achieve the desired result.

2.3 PROBLEM DEFINITION AND MODELING

One of the common consumer behaviors is brand switching or switching costs. This buying behavior of customers causes them to avoid buying a product with a regular brand and replace it with another product or brand. Now, the purpose of this research is to examine the relationship characteristics, switching costs, and customer characteristics. The research findings show that there is a significant relationship between the duration of receiving any service to customers and the type of customer purchase, with switching costs. Also, a good experience of purchasing products and receiving services from a store or supplier of goods and services influences the relationship between the duration of service received by customers and the type of customer purchase with switching costs. Finally, the willingness of customers to switch stores and the provision of services from a store (or supplier of goods and services) moderate the relationship between the duration of receiving any service to customers and the type of customer purchase with switching costs.

Now it can be said that the more attention companies pay to customer behavioral characteristics, the more loyal customers will be to the company's products and the possibility of changing stores or, in other words, imposing switching costs and creating losses for the company will be reduced. Types of switching costs include: transaction costs, equipment compatibility, learning to use new services or equipment, uncertainty about the quality of new services, and search costs for new suppliers or customers. These issues make it more expensive for companies to attract new customers than to retain current customers, so customer retention is a more appropriate strategy for generating profits. This research seeks to search for the most influential factors through the Levenshtein distance algorithm and the PPM approach. The PPM framework is divided into three parts: Push Effects, Mooring Effects, and Pull Effects. The Levenshtein algorithm is also an implementation solution based on distance, which will be briefly explained in this section.

The main purpose of using social network analysis methods is to extract complex information. Software in this field has three features:

- Community identification
- Network level calculation
- Graphical visualization of the network

One of the most common network analysis metrics is the centrality metric. This metric measures the importance of a node or a node in the entire network. Degree centrality, betweenness centrality, and closeness centrality are the most widely used examples of centrality metrics. The degree centrality of a node is the number of nodes that are directly adjacent to the node, so the centrality of a node indicates the greater importance of the node in the network. In fact, this model calculates how many nodes in the network need this node to communicate faster. Closeness is the inverse of the average distance of a node from other nodes in the network. The node with the highest degree of proximity has faster access to other nodes and can receive or send information from all nodes in a short time.

There are two other important metrics for measuring the network, which are: density and clustering coefficient. Network density can be defined as the number of existing relationships relative to the number of possible ones. The clustering coefficient performs a different function than the centrality measure and is more similar to the network density measure. In many network analysis models, only the existence of a relationship is important, and the direction of the relationship is not important, which is what we call an undirected network. In other words, an undirected network has a set of nodes that are connected to each other. These discussions were a prelude to the basic discussions of networks and are relevant to our discussion in the field of social network analysis, which we will explain in the following. These include the regression method, the PPM model, and the Levenshtein distance algorithm.

2.3.1 PPM Model

2.3.1.1 Push Effects

The first P in the PPM method examines the push factors in customer switching. Push effects refer to negative factors that make customers dissatisfied with their current provider and encourage people to leave their current situation (change service provider). If we were to examine these factors, we would come to the following:

1. Weakness in after-sales service
   After-sales service is a major reason for further interaction with the customer. As long as the after-sales service is considered perfect for the customer, the customer will go to your brand with more hope. Customers dissatisfied with after-sales service will never express this dissatisfaction. Instead, they will

look for a brand with better service. The reason for this is quite clear. A brand that has poor support services cannot perform well in other areas. In studies to discover such customers, keep in mind that zero dissatisfaction can also be dangerous. If no customer has reported dissatisfaction due to poor performance, it indicates that the customer has left you without complaint or has not found a way to express it.

2. Customer communication channels
   Customer churn does not always indicate dissatisfaction; perhaps this dissatisfaction has not reached you. So, do not always make the level of dissatisfaction the criterion for the performance of your services. First, make it easy for customers to communicate, so that they can contact you in the shortest time and express their level of satisfaction.
3. Customer expectations
   Is your product really what you are advertising? Can it satisfy the customer? Do you know the experiences of your customers after the purchase? After a bad experience with the product and not receiving good support services, a dissatisfied customer will share this level of dissatisfaction with their friends, and thus other customers will be discouraged. Know your customers' needs before selling, and do not seek to appear more than expected. A quality product will always have its fans. Many problems arise after the sale without testing. Always deliver a tested product to the customer.
4. Value of the product to customers
   The value of a product is determined after purchasing it. In this way, the customer, considering the amount he paid for the price of the product and the function of the product, concludes that this product is worth this amount, and this does not necessarily mean a low price for a product. The price of the product is always evaluated equally with its value, for example, considering the function of Apple phones, a person accepts that buying a phone from this brand is worth it and will not regret it after buying. The unique features of a product are the value that the customer is looking for. In this way, after paying the amount, he can look positively at the high value of the product he has purchased. When your product has a special feature that increases its purchase value, announce it in your advertising. You can even provide it to the customer on a trial basis so that the value of the purchase is truly determined for them.
5. Distrust
   A customer can remain loyal to your brand when their trust has been earned. Trust is also gained through consistency in producing quality products and excellent services. Your customer is no longer worried about buying from you. The promises made in advertising or the services that the customer expected to be fulfilled must be realistic, because such advertising will no longer have an effect on anyone after some time.

2.3.1.2 Mooring Effects

The second part of the PPM framework is the Mooring Effects section. If we were to describe this section in one sentence, we could say a situation where the customer is in a state of rest and does not make any changes in their purchasing policy. In other words, the customer does not have an urgent need to buy or has not yet made a decision to buy. To change the customer's status from this state, the following can be used:

1. Discount
   For a customer who is in a state of rest, a discount can be a good motivator. A discount always attracts attention, because they see it as an opportunity. For people who have just started their business or do not have enough customers, a discount can be a good means of getting recognized. If what you offer is of good quality, you have a chance that they will come back to you. Set the discount to a level that does not make your product or service look worthless, and on the other hand, the discount should be at a level that encourages the buyer to buy. Every purchase is considered a risk for the buyer, and you intend to reduce this risk for them with a discount. Sometimes a discount encourages the customer to buy more because they know that they will not have such an opportunity tomorrow.
2. Targeted advertising
   What makes the buyer decide whether to make a purchase or not based on an advertisement in physical or online sales is the skill of advertising. Advertising writing is a very important skill in sales. Place advertisements where your customers are. A good advertisement, especially if it is seen several times, will make you more memorable in the minds of potential customers. Customers need your services, but one issue remains: how do you introduce yourself to them in advertising? In advertising, you do not talk about how good you are. Focus only on one thing: what added value you add to the customer's life.

2.3.1.3 Pull Effects

The last part of PPM is pulling factors. Pull factors make the customer tend to lean toward the products or services offered by us. The results show that all PPM factors affect customer intention, and factors related to current service providers (push factors) have less effect than attracting other people (pull factors). If we were to examine these factors, we would come to the following:

1. Sell the customer's needs
   Always keep in mind the premise that customers buy what they think they need. How can you convince them that they need your products? Highlight the benefits of the product that reduce costs, time, and solve problems. Product features such as weight, color, and size are not the main drivers of purchase. Show the buyer what they will be able to do more easily if they buy? What will make them feel better? What value will be added to their life? What positive change will it make in their lifestyle? How will the product save the customer time and money?
2. Focus on a specific segment of the market
   Few businesses have the resources, time, and budget to focus on a large segment of the market. Focus on a specific range of customers to become a brand in this smaller market. You can't write an ad that will appeal to

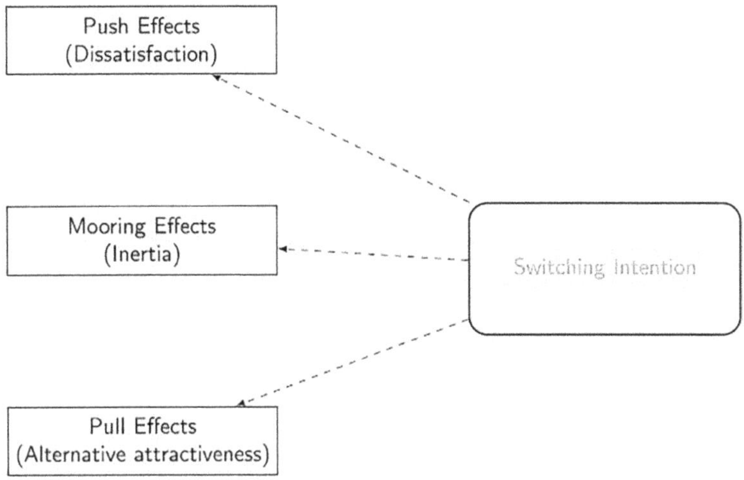

FIGURE 2.1 PPM components.

everyone, nor can you have a product or service that will appeal to everyone. By focusing on a segment of the market, your customers are more targeted, and you can sell more to them.

3. Satisfying existing customers
   Instead of just thinking about attracting new customers, it is better to try to satisfy existing customers as much as possible. Retaining existing customers is much more important than attracting new customers, and it costs us less. Results show that blogger satisfaction (push factor) is the most important factor in changing behavior, followed by the attractiveness of alternatives (pull factor) related to other features and capabilities offered by competitors and the costs of switching (mooring effects factor).

A configuration of the PPM model is shown in Figure 2.1.

2.3.2 Regression and Classification

Logistic regression is a statistical method that estimates the relationship between input and output variables and is widely used in statistical modeling, where the relationship between a dependent variable and one or more independent variables is modeled. Regression can effectively model dichotomous outcomes and has the ability to assess the relationship between dependent and independent variables numerically. Logistic regression models can be binary or multinomial. The values of the dependent variables are usually coded as zero and one in binary logistic regression models to represent two possible outcomes. In relation to customer churn, it is modeled as a function of various variables as follows: If churn occurs, $Y = 1$, and otherwise $Y = 0$.

Specifically, two conditional probabilities are $p(y = 1|X)$ and $p(y = 0|X)$ and $X = (X1, X2, \ldots, Xn)$, where n represents the factor associated with customer churn (independent variables). The output of the following model ultimately examines the probability of customer switching to another store, and its goal is to minimize the customer classification error.

$$\log\frac{p(x)}{1-p(x)} = \beta_0 + \beta_1 X_1 + \beta_2 X_2 + \cdots + \beta_n X_n \tag{2.1}$$

In this formula, P represents the percentage of customer churn based on switching. Also, β_i represents the correlation coefficient, and Xi is also determined as part of the logistic regression analysis. Of course, this formula or model can be rewritten as follows:

$$P(Y \mid X) = \frac{1}{1 + exp(-(\beta_0 + \beta_1 X_1 + \beta_2 X_2 + \ldots + \beta_n X_n))} \tag{2.2}$$

In this formula, we set a threshold delta for switching. That is, when we set this delta and $\Delta < p(y = 1|X)$, we conclude that the customer is likely to remain loyal to us and does not intend to switch. Logistic regression has the ability to identify independent variables that are statistically significant in affecting the outcome (customer switching). Logistic regression analysis is usually performed on raw data after applying appropriate data processing (cleaning, balancing, transformation, etc.) to achieve meaningful results. The statistical method of the R software in the regression section used to predict customer churn is as follows:

$$\text{P(Churn)} = C - 0.075X_1 - 0.633X_2 + \cdots + 4.850X_{18} \tag{2.3}$$

where C is a constant, and Xi represents the independent variables in the system. The coefficient of each independent variable indicates the relative influence of the variable. For example, if X1 increases as one of the factors, using this unit, the probability of customer churn decreases.

While researching the causes of customer churn is an important issue, it is also important for researchers in this field to understand the behaviors of potential customers and other customers. This can help them to tailor strategies to attract other customers and potential and loyal customers.

For this analysis, we can consider the dependencies between two customer factors. The usual choice for inferring the dependency between two factors is the Pearson correlation coefficient.

The ρ_p between two random variables X and Y is defined as follows:

$$\rho_p = \frac{E[X \times Y] - E[X] \times E[Y]}{\sigma_X \times \sigma_Y} \tag{2.4}$$

In this formula, σ_X and σ_Y are the standard deviations of X and Y, respectively. While this measure of dependence is widely accepted, it has limitations in quantifying the degree of association between random factors. A suitable measure of dependence should have eight properties, all of which ρ_p does not satisfy. For example, ρ_p violates invariance. ρ_p is a linear measure and is useful when dealing with a linear relationship. In other words, ρ_p is not suitable when the linearity of the relationship is not guaranteed. So, we need a different method that can take into account the Spearman (ρ_s) and Kendall τ correlation measures to quantify the association between two factors in customer data. While either measure can be used, we prefer Spearman (ρ_s) because it is less computationally expensive to compute than Kendall.

Spearman (ρ_s) is also calculated as:

$$\rho_s = 1 - \frac{6\sum d_i^2}{n^3 - n} \tag{2.5}$$

where di is the distance between the two random variables, and n is the number of data points. For customer classification, which is an important topic in our discussion, the following modeling can be done. The attempt is to build a suitable model or classifier that can classify everything. To evaluate and compare the performance of all the prediction models of arousal with the relevant parameters, several performance measures such as the number of positives (TP), false positives (FP), true negatives (TN), and false negatives (FN) are used in this study. The confusion matrix includes the predicted results and the actual outputs of the prediction model.

In this section, we use three performance measures to compare the models: sensitivity, specificity, and accuracy. The higher the value of these measures, the better the performance of the prediction model.

1. Sensitivity
   It deals with the probability of correctly predicting the customers who will complete their services with the company. When the sensitivity value is high, the model's ability to predict customers who are likely to have high arousal is high. Sensitivity can be calculated as follows:

$$Accuracy = \frac{TP + TN}{P + N} \tag{2.6}$$

2. Specificity
   In this criterion, the probability of correctly predicting non-loyal customers who are likely to churn is high. When the value of the feature is high, the ability of the model to predict non-loyal (churning) customers is high. This criterion can be calculated as follows:

$$Specificity = \frac{TN}{TN + FP} = \frac{TN}{N} \tag{2.7}$$

3. Accuracy
   The overall accuracy of classifying each customer so that the type of customer (loyal or churn) is done with high accuracy is calculated as follows:

$$Accuracy = \frac{TP + TN}{P + N} \tag{2.8}$$

2.3.3 Levenshtein Algorithm

Regarding the Levenshtein algorithm, which is another algorithm we use, it can be said that the Levenshtein distance algorithm is widely used to measure the similarity of two words, which is also used in spell checking and autocorrection. In this method, the edit distance is calculated to measure the similarity between two words.

The edit distance is defined as the minimum number of edits required to convert an incorrect word into a correct word.

In this research, we used the Levenshtein distance algorithm to classify the data of the PPM method and then analyzed the accuracy and effectiveness of different models using statistical software with the resulting output. We used the Levenshtein (LV) algorithm to measure the distance between two sequences of order Si and Sj. C is the local cost matrix that shows all pairwise distances between Si and Sj.

$$\begin{aligned} LV(Si, Sj) = \min\{&LV(Si-1, Sj-1), LV(Si-1, Sj), \\ &LV(Si, Sj-1)\} + C(Si, Sj)\} \end{aligned}$$

We first collect the data, then analyze them, and sort and balance the data. In the next step, using the predictive models that are used extensively in this research from the PPM framework, we then examine and draw conclusions from the output. A proposed flowchart based on the research stages is shown in Figure 2.2.

2.4 PERFORMANCE ANALYSIS

Table 2.1 shows the measured values for training and testing the correct detection rate of each method. Regression, SVM, and Random Forest have more than 90% accuracy for both training and testing sets. The important factors identified to influence customer switching behavior are gender, number of children, schedules, number of phones, consecutive month of income, percentage of decline, and age. The probability of switching behavior for a particular customer can be obtained by using a logistic regression model.

The logistic regression, SVM, and random forest models outperform the decision-tree model in all performance measures, as shown in Table 2.1. Three important factors that influence customer switching behavior on balanced and unbalanced data are discussed in Table 2.2. As expected, the analysis performed on the balanced dataset is more accurate. The results obtained on the unbalanced dataset can be misleading due to the low ability to learn from the minority class. Therefore, the analysis performed on the balanced dataset provides a more accurate prediction of customer switching behavior.

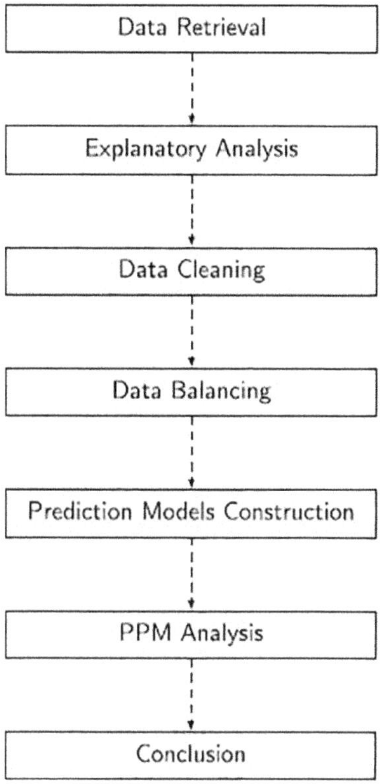

FIGURE 2.2 Research flowchart.

TABLE 2.1
Measurement of the Prediction Models

Decision Tree	Random Forest	SVM	Regression	Performance Measurement	Stage
83.11	87.62	88.33	89.02	Sensitivity	Preprocessing
90.03	95.07	94.45	93.13	Specificity	Preprocessing
86.66	91.34	91.38	91.07	Accuracy	Preprocessing
83.99	88.22	88.47	88.77	Sensitivity	Test
89.75	94.38	94.66	93.59	Specificity	Test
83.11	87.62	88.33	89.02	Accuracy	Test

Accordingly, we will use the balanced dataset for further analysis, drawing insights and conclusions. The top three features of the logistic regression model in the sections are the 15-second dropout rate, whether a customer is still under contract, and the number of phone calls. The 15-second dropout rate means the percentage of calls that are dropped within 15 seconds. In other words, when a customer contacts the

TABLE 2.2
Three Important Factors

Data Type	Regression	Decision Tree	Random Forest
Unbalanced Data	Drop 15%	Drop 15%	Drop 15%
Unbalanced Data	Under Contract	Consecutive Months	Minutes
Unbalanced Data	Percent Data	Minutes	Revenue
Balanced Data	Drop 15%	Drop 15%	Drop 15%
Balanced Data	Under Contract	Consecutive Months	Consecutive Months
Balanced Data	Service Call	Minutes	Under Contract

company and fails to make a purchase after 15 seconds, and the company fails to persuade the customer in the first seconds of the call. The number of service calls means the number of calls that a customer makes, for example, to request assistance from the R&D department. If a customer experiences a higher number of phone calls within 15 seconds and needs more service calls, this means that there is a higher chance for this customer to switch to other competitors. However, if the customer is under contract, there is a lower chance that the customer will switch to other competitors.

In addition to having high sensitivity and accuracy, logistic regression is the easiest algorithm to interpret, so it can be widely used to predict customer switching behavior. The results of the random forest show that the percentage of 15 dropouts, the sequence of months, and whether the customer is still under contract are the most important factors that affect customer switching behavior. The "consecutive months" factor means how long a customer stays with the vendor and is loyal to it. The percentage of drop or decrease of 15 is still the most important factor, and being under contract with the customer is also in the top three factors, but instead of the number of service calls, consecutive months become an important factor. This shows that the longer a customer is with a vendor and the vendor maintains their service, the less likely the customer is to switch. The decision-tree results tell us that the percentage drop of 15 consecutive months and minutes are the most influential factors on customer switching behavior. The minutes factor here means the average time of contact with the customer. This means that if the average contact of a customer is higher, the customer is less likely to switch. Overall, the 15 percent drop is a factor that affects customer switching behavior in all models, whether the data used for the analysis is balanced or unbalanced.

2.4.1 Analysis

In this research, we have reached the mentioned influential factors by conducting field and research studies and analyzing the factors affecting customer switching behavior in Iranian and international online sales companies, as well as local businesses, and finally by ranking the resulting assumptions, which we have reached at this stage using the data obtained and the statistical software. The analyses conducted so far do not distinguish between factors that push customers to other factors (push factors) and factors that make customers hesitate to create switching behavior (mooring factors). For example, the 15% reduction factor is a common type of push, because a customer may be inclined to another seller if they experience a large number of phone calls. The contract factor is an excellent deterrent that can be dictated to customers, and if they show switching behavior before the contract ends, they face a penalty.

To further study and analyze the factors influencing customer switching behavior, we divide the features in the data into two categories: push and restraint, as shown in Figure 2.2. Then, we combine the PLS-DA and PCA methods and perform a stepwise analysis (PCADA) by combining the factors of these two methods. Note that there is no pull factor in the data, and therefore we perform a scenario-based sensitivity analysis to examine the effect of pull on customers' intention to switch behavior. We first perform the push analysis, excluding all factors except the push factors. Then, we repeat the same procedure for the mooring analysis, the pull analysis, the push-mooring analysis, and finally the push-pull-mooring analysis. The PCA-DA and PLS-DA analyses are performed using statistical data. The R software was trained on a training data set based on customer switching behavior with 70,212 case observations and 18 variables, providing an explanation for predicting customer switching behavior. We use the average accuracy for the analysis. When there is no significant difference in the performance of the models, the simplest model should be selected. A standard error (in itself) selects at least one complex model with a cross-average accuracy of the optimal model. In addition, a range of values should be defined for the tuning parameter, which specifies the number of latent variables (LVs), i.e., the number of components. In this section, we consider all models with 1–20 LVs (i.e., components).

The analyses are performed, and then the results are compared. It follows from the results that the accuracy of PCA models is lower than that of PLS models. For example, the accuracy of PPM25PLS is 93% while the accuracy of PPM25PCA is 90%. Therefore, we will use PLS further in the following. Also, the model that shows the best performance and highest accuracy is PPM75PLS with 98%, and the Mooring PCA model with the lowest accuracy is comparable with the accuracy number of 77%. It is clear that higher accuracy can be achieved by using more factors. Moreover, the accuracy of PM (Push-Mooring) is exactly similar to PPM, with 0% of customers affected by the pull effects. Theoretically, PPM with 0% of customers affected by the pull effects should be like PM, which is confirmed by this result. Therefore, only PM is used in the subsequent analysis.

Finally, the 10 features that affect customer switching behavior are obtained by using the variable importance in the image from the PLS-DA model.

2.4.1.1 PLS-DA

A PLS-DA method for push features is performed to discover the most important push features that affect customer switching behavior, without any concern about conflict with other factors.

The accuracy of the PLS-DA push method is 80.06%. The most important push feature is the percentage drop of 15, followed by the number of service calls. Note that in this section, three factors are considered: percentage drop of 15, number of service calls, and bill factor.

2.4.1.2 Mooring PLS-DA

The Mooring PLS-DA analysis is performed to discover the most important mooring features (stationary state) that influence the customer switching behavior. The most important feature in the Mooring discussion is the number of consecutive months (number of consecutive months that the customer has been using the vendor), minutes (average call time), data percentage (amount of data used in percentage), and age (customer age). As expected, being under contract with the customer was the most important factor. However, it is somewhat surprising that consecutive months are much more influential than minutes and data percentage, since the latter two are directly related to the amount of calls and data used by the customer.

Consecutive months have no relation to actual usage. One interpretation is that consecutive months may be a good measure of how loyal a customer is to a vendor. Note that age is the fourth factor, which indicates that the older a customer is, the less likely they are to switch. This is actually more consistent with our common sense because psychologically, people tend to switch less as they get older. A push-mooring PLS-DA analysis is performed to discover the most important factors, outside of push and mooring factors, that influence customer switching behavior. The accuracy of this method is 90.26%.

The most important feature is the percentage of dropout at 15, followed by the features under contract, consecutive months, minutes, age, percentage of data, and revenue, in that order. Note that the percentage of data now follows age (which is in the reverse direction of the mooring analysis); however, the difference between the two is very small. Also note that the Bill factor is not in the top 10 factors. In this Push-Mooring analysis, we conclude that the customer is much more sensitive (more than 25%) to the factor of 15% drop vs. the factor of being under contract. It can also be concluded that the accuracy is higher when more factors are included in the model, and this increase in the number of factors helps us to have a more accurate model. pm analysis (pm PLS-DA) with 90.26% accuracy, push analysis (push PLS-DA) with 80.6% accuracy, and mooring analysis (mooring PLS-DA) with 78% accuracy are ranked first to third based on the accuracy parameter.

2.4.1.3 Push-Pull-Mooring Analysis

As there is no pull factor in the data, we create three factors: better service than other companies, better reputation than other vendors, and peer influence. These factors are created to simulate the effects of pull and study how they can influence customer switching behavior, along with push and inhibit factors. Similar pull factors are also found in research. After the three pull factors are determined, a sensitivity analysis is performed using the developed scenarios. In the first scenario, we assume that 25% of switching behaviors are affected by all three push factors. More precisely, we randomly assign the first pull factor, which is better service, with a weight of 25%. Then, we start randomly and assign a better reputation with a weight of 25%. Of course, the second 25% is completely independent of the first 25%. We repeat this process for the third factor, which is peer influence. This creates the first scenario. Then, we repeat the same process to create the second scenario (50% influence) and the third scenario (75% influence).

In each of these three scenarios, the three factors exerting pressure on switching behavior are the same. PLS-DA analysis is performed using three scenarios: 25%, 50%, and 75% of switching behaviors are affected by the pull effects of other competitors. In scenario 1 (25% of switching behaviors are affected by the pull effects), the accuracy of the PLS-DA model is 93%. The top predictive feature is the percentage drop of 15, followed by the features under contract, peer influence, better service, consecutive months, and better reputation. Using scenario 1, we may conclude that the relative importance of the consecutive months factor is about 25% because it is close to 25% of the pull factors. We may also conclude that the percentage drop of 15 and under contract is more important than 25%.

In the second scenario, the results of the accuracy of the PLS-DA model is 96%, and the top predictor features are 15% drop, better reputation, peer influence, better service, and under contract, respectively. And finally, in the third scenario, the results of the accuracy of the PLS-DA model is 98%, and the top predictor features are better service, better reputation, peer influence, 15% drop, and under contract, respectively. Overall, we may conclude that 15% drop is more than 50% and less than 75% effective, and under contract is the second most influential factor, with an importance of less than 50%.

To analyze the negative aspects of customer switching behavior and analyze how pull, push, and anchor factors can influence each other, we used a scenario-based approach. In particular, pull factors are carefully designed and developed to be incorporated into our scenario-based methods. Due to our novel approach, we are able to gain new insights, which were discussed in detail above.

This is a very unique technical approach, and we believe that our book is the first to use such an approach. Because of this, we can obtain results on predicting customer switching behavior and study their change as pull factors, trying to move away from the purely theoretical aspect and match the subject with real-life factors.

2.4.2 Dependency Analysis

In this section, we perform dependency analysis to better understand customer switching and non-switching behaviors. For this, we use Spearman's rank correlation (ρ_s). For each set of these records, Spearman's correlation (ρ_s) was calculated pairwise, and the results are computed, where each factor is coded as follows: F01 (consecutive months), F02 (calls), F03 (income), F04 (bill), F05 (minutes of phone talk), F06 (data percentage), F07 (drop percentage 15), F08 (age), F09 (children), F10 (education), F11 (number of phones owned by the customer), F12 (number of phone lines), F13 (service calls), F14 (gender, male = 0, female = 1), and F15 (contracted, yes = 1, no = 0). In particular, customers who are not inclined to switch are more likely to stay with the service provider even after the contract ends, so the ρ_s of F15 (under contract, yes = 1, no = 0) and F01 (consecutive months) is equal to −0.84. This shows that the more non-switching behavior a customer has, the less likely it is that they will be under contract with the service provider. However, the ρ_s of switching-prone customers is equal to –0.09, indicating that there is no strong correlation between F01 and F15 in the behavior of switchers. This is an unexpected result, since customers with switching behavior would be expected to switch to another service provider after the contract ends, which turns out not to be the case. This could suggest that instead of offering promotions to retain customers, providers should focus more on service quality. Another interesting comparison for F05 (number of minutes) and F07 (percentage of drop 15), ρ_s between F05 and F07 for customers with switching behavior is –0.71 and for customers without switching behavior is –0.56. It can be concluded that customers with switching behavior talk and comment more when the service quality is good, and they are customers with more weight (buy more) than customers without switching behavior. In the comparison between F07 and F13 (call service), interesting behaviors are also created regarding service quality, ρ_s for switchers is –0.21 but for customers not inclined to switch is equal to 0.

This clearly shows that switchers care a lot about service quality, because switchers use call services more and comment on the service, and this position has caused the service quality to improve, and this behavior was not observed from customers not inclined to switch. In other words, even if they make service calls, the quality of service does not improve because they are not interested in commenting and act indifferently. Therefore, the correct interpretation could be that customers who are inclined to switch care much more about quality. Also, customers who are inclined to switch use service calls much more when they are under contract compared to customers who are not inclined to switch. ρ_s between F13 and F15 shows this, ρ_s for customers who are inclined to switch is equal to 0.18 and for customers who are not inclined to switch is equal to 0.01. Finally, according to the results obtained, the characteristics of a 15% drop, consecutive months, and being under contract can be identified as the most important factors.

2.5 CONCLUSIONS

As one of the most important problems and threats that can of course become opportunities, the issue of retaining and maintaining current customers is the one we addressed in this research. The goal we pursued was to identify the factors that had the greatest impact on customer switching behavior. We tried to adopt an appropriate approach by identifying algorithms and examining the factors involved in this issue and helping to improve the online business environment and even traditional businesses. Companies and individuals can use the method and results in this chapter to help make effective marketing activities and strategies more efficient for customer retention. The statistical methods and models used in this chapter before the PPM analysis that was the basis of our work are logistic regression, SVM, random forest, and decision-tree methods. A future and developmental suggestion of this research could include examining repulsive and effective factors that create more realistic scenarios for PPM. In addition, the use of other data mining methods in building predictive models, such as artificial neural networks and Bayesian networks, has the potential to increase the scope of this research. It is also possible to use hybrid data mining methods and make various comparisons with the predictive models of this research. Finally, it is suggested that based on this research, sales and marketing companies cooperate with companies active in the field of statistics, and through this cooperation, companies related to the provision of services and products can keep pace with customer trends in the customer switching process and improve customer retention strategies, incentive programs, and marketing activities. This cooperation also gives us a good opportunity to learn from professionals and improve our future research.

3 Clustering and Automatic Cell Diagram for Customer Segmentation

3.1 INTRODUCTION

E-commerce is a term used for trade through information and communication systems. Although e-commerce has not been around for a long time, it has assumed a significant role in our daily lives, so that it is not easy to be avoided. One of the simplest and most effective roles of e-commerce in the daily lives of Internet users is the purchase and sale of goods and the exchange of related funds through smart cards.

Electronic commerce is a set of communications, information management, and security capabilities that allow organizations, businesses, companies, the public, government, etc. to offer their services and goods more efficiently, more easily, and more quickly using computer communication networks, especially the Internet. Electronic commerce is paperless commerce. Through electronic commerce, the exchange of information, buying and selling, and the transportation of goods are done with less effort. Internet commerce makes communication between companies, businesses, and all institutions related to trade easier and faster. Electronic commerce is a set of activities and services that make it possible to search for information, manage information and exchanges, check credit status, create credit, make virtual payments, report, and manage credits on the Internet with digital communication. Many people consider electronic commerce to be limited to buying and selling via the Internet, while this is only a small part of electronic commerce. In principle, electronic commerce can encompass all aspects of business and economics. Simply put, any commercial and financial activity between different institutions and individuals can be considered e-commerce. And it can be said that e-commerce is an electronic bridge between commercial centers, buyers, and sellers. The purpose of e-commerce, which is actually a combination of commercial techniques and methods, is to cooperate, compete, and increase the efficiency of different companies, so that over the past 15 years, it has enabled companies to improve their business methods and improve their communication with related companies, and finally expand the scope of commercial activities, even on a large and global scale (Shahri et al., 2013).

DOI: 10.1201/9781003740148-3

3.2 PROBLEM STATEMENT

In today's world, there is no need to drive to the store and park, wait in line for the bus, and argue with sellers. In some cases, they can even set their own asking prices (e.g., in the case of hotels and airline tickets) so that suppliers who can meet the buyer's requirements can contact them, in other words, make a bid. Buyers can benefit from greater discounts by shaking hands with others and increasing their purchase volume. In today's world, buyers can order almost anything they want via the Internet (in our country, the path of developing online trade and business tends to this goal), such as home furnishings, washing machines, management and legal consulting, and medicine.

Online shopping is an integrated system that includes negotiation, ordering, receiving, and post-purchase review operations. Understanding the needs and expectations of existing and potential customers is essential to maintaining business success in the market. Reaching the largest number of online shoppers is only possible by understanding their behavior, which includes all the activities involved in acquiring and subsequently using and distributing products and services, along with their prior decisions and conditioning. Taking into account customer profile information, it is possible to customize the offer and information on websites based on their preferences (Fazlollahtabar, 2022a).

In the current electronic age, most organizations have established an online presence. This presence ranges from promoting the image of the organization to introducing its products and services to the possibility of selling its products and services online. On the one hand, the rapid growth and diffusion of Internet technology have increased the popularity of visiting websites among contemporary visitors, and on the other hand, recent marketing strategies consider customers as important resources of an organization. In fact, with the transition from a traditional economy and the increase in competition in new dimensions, the customer is the main pillar and the focus of all company activities. In a way, the survival and continuity of the life of organizations depend on identifying and attracting new customers and retaining existing customers. Therefore, gaining knowledge about customers and understanding their needs are very essential to retain customers in an online e-commerce, because competitors are just a click away from us. Predicting the buying behavior of online customers is difficult because their visits to stores rarely lead to actual purchases, which has become a challenge for marketers and researchers.

The average visit-to-purchase conversion rate for online businesses is at most 3%. In this case, a helpful way is for organizations to target only those customers who have achieved certain profitability criteria based on their individual needs or purchasing behaviors. In fact, this greatly contributes to the most important marketing goal, which is to know customers and understand their needs and, as a result, to make more sales. Therefore, there is a need to collect and analyze customer click behaviors on the website, because attracting new customers and retaining valuable customers are of great importance. Web scraping is an important step in achieving this goal.

3.2.1 Objectives

Due to the high dependence of e-commerce on direct interaction with the customer, the issue of establishing appropriate customer service is one of the most important issues that companies and suppliers must consider, including activities related to product recommendations and online advertising. In addition to being beneficial to sellers, e-commerce is also beneficial to customers. Customers can benefit more from e-commerce by spending less effort to find products and services, at the lowest price, and with up-to-date and new goods. Another benefit is that online stores are open 24 hours a day and provide business services to customers 7 days a week.

3.2.2 Assumptions

The subject area of this research is better customer relationship management, examining customer purchasing behavior models, and then providing suggestions such as dedicated advertising, expressing customer favorite products, and guiding them to choose the right supplier.

Paying attention to each customer's shopping process and recording their activities, including clicks, searches, reading catalogs, etc., one can review and analyze and then personalize the shopping environment according to their needs and preference, in order to create a special interest and connection with the customer, and provide them with the shopping process in an optimal and timely manner.

3.3 BACKGROUND

In 1991, scientists first founded e-commerce, and in 1994, when online banking was created, researchers started their first online sales. In the same year, Netscape invented and created Secure Socket Layer (SSL), and online shopping was created using this protocol. Also, in the same year, the German company Intershop created its first online sales system. In 1995, Amazon.com and (in 1996) eBay also started their activities. The e-commerce environment, which is based on the Internet, allows customers to search for information and purchase goods and services through direct communication with online stores. It should be noted that online shopping is not based on the actual experience of buying goods, but on appearances such as images, shapes, qualitative information, and advertisements of goods. In the works and articles related to e-commerce, there are numerous studies on online shopping and consumer evaluation, some of which justify the existence of an empirical validity between the factors affecting shopping and online shopping behavior. However, new research has considered the information processing style of customers, shopping patterns, customer preferences, and related fields to create more attractiveness for users of online stores. Marketing researchers believe that behind every purchase, there is an important decision-making process that needs to be examined. The steps that the buyer goes through to decide what type of products to buy are called the purchase decision process (Fazlollahtabar et al., 2012).

The word service includes various meanings. A service is a type of economic activity that creates value and provides benefits to customers by resulting in a desired change in the recipient of the service or on their behalf at specific places and times.

A service is a result that customers want. A service is an activity or benefit that one party offers to another that is essentially intangible and does not involve ownership of anything. Services are intangible and volatile products that are produced and consumed simultaneously. Quality is the degree to which a customer's needs are met, which is related to organizational culture. Quality is a set of activities, processes, practices, and interactions that are provided to customers in order to solve their problems. A product is of quality when it meets the customers' needs and wants. Kelce believes that service quality includes three dimensions: physical, situational, and behavioral. In other words, service quality is a focus on what is delivered to customers, the situation in which the service is provided, and how the service is provided. Service quality is the organization's continuous attention to meeting the needs and wants of its customers.

Electronic satisfaction refers to a customer's satisfaction with a previous purchase experience from an electronic business. Electronic satisfaction refers to the customer's satisfaction with the site design, information, or content of the site, shopping convenience, and shopping security. Electronic satisfaction refers to the preference of an online store's products or services over competitors' products or services when making a purchase. Electronic satisfaction refers to the level of customer satisfaction with support for receiving and sending orders for products or services, after-sales service, price of products or services, quality of website content, website speed, website reliability, website ease of use, and security (financial, privacy).

In a study titled "The Effect of Consumer Beliefs on Impulse Buying Behavior in the Online Environment", the aim of this study was to investigate the effect of online shopping beliefs on impulse buying behavior of consumers in online shopping. Finally, the results of the study indicate the significant effect of enjoyment during search on purchase motivation and the significant effect of purchase motivation on impulse buying; therefore, to stimulate purchase, online retailers should create a relaxed, friendly, and entertaining site with an attractive layout.

In another study titled "Investigating the Factors Affecting Online Shopping: Explaining the Variables of Trust, Perceived Risk, Perceived Benefit and Their Antecedents", the purpose of this study is to investigate the factors affecting online shopping, identify and influence the antecedents based on cognition, experience, emotional state, and personality on these factors among customers of computer websites. Based on the results obtained from linear regression analysis with SPSS, all research hypotheses have been confirmed, so that the consumer's purchase intention has a positive effect on their purchase decision (Fazlollahtabar and Ashoori, 2020).

In 2016, a study titled "Modeling Online Consumer Behavior: The Superior Role of Emotions and Moderating Effects of Need for Cognition and Optimal Level of Arousal" reached these results, which, in the context of studying the relationships of the aforementioned elements, showed that in the context of online shopping, emotions play a positive role before cognition. Among other findings, the dimension of mastery of information and cognition played a major role in the model. The moderating effects of the optimal level of arousal and involvement with shopping were examined separately, and their interaction effects were also tested.

Despite the global rise of e-commerce, Malaysia fell from 30th to 10th place in the global retail e-commerce rankings in 2015. Malaysia's decline in spending power is in line with the global financial crisis. For a middle-income economy like Malaysia, it makes sense for citizens to limit their spending, especially when Malaysia has recently experienced a rapid decline since 2013. Therefore, this situation has not created any problems for several industries, including online commerce and foreign direct investment in our country. The global economic crisis has forced companies and entrepreneurs to turn to online marketing as the cheapest way to advertise and reach a customer base in the short term. Masaya Ono, Vice President and CEO of Rakuten Online Shopping, found that online shopping in Malaysia is gaining traction. Recently opened web stores such as 11 Street, a brand from South Korea, can be seen as a forbidden attraction in the competitive Malaysian home e-commerce market.

Another local retailer that is online this year is Atoz, which sells electronics, stationery, groceries, and other goods. In fact, e-commerce in Iran is one of the latest examples of this, and it is important to examine the factors that influence online shopping behavior.

Changes in consumer behavior over the decades have led retailers to delve into the psychology of virtual shoppers. The problem of deciphering online customer behavior becomes more complex with zero interaction during transactions.

The huge investment in predicting virtual shoppers has forced online retailers to pay more attention to understanding customer behavior. Although both the government and the private sectors have invested heavily in promoting online shopping, brick-and-mortar stores are still the instinctive choice for the majority of customers. A 2012 study by the Open Thinking Exchange IPOS found that 56% of respondents from 24 countries prefer to shop in brick-and-mortar stores to online shopping. Even in developed countries such as the United States, 63% of surveyed consumers search online before making a traditional purchase, but only half of consumers actually purchase online. The ability to change behavior and cultural constraints is another factor that prevents Malaysians from shopping online. Consumers tend to listen to word-of-mouth recommendations from close family and relatives, friends, or even the media to make purchasing decisions. Therefore, to change Malaysian consumer perceptions of online shopping, online retailers must make great efforts in promoting and implementing good customer service to increase interest in online shopping. A study by Rakuten in 2010 found that 71% of Malaysian online shoppers regret their online purchases, 48% are dissatisfied due to unmet expectations, 29% are disappointed with poor product quality, and 30% of Malaysians are unable to complete an online purchase. Customers also find it inconvenient when they encounter problems with payment, limited product information provided, and the online retailer cannot be reached by phone (Fazlollahtabar et al., 2013).

Retail e-commerce grew to approximately US$840 billion in 2014, up from US$695 billion in 2013, and is estimated to grow to US$1,506 billion by 2018. The continued growth in sales suggests that e-commerce has enormous market potential. Online retailers should design their websites in a way that makes them appear customer-centric and actively manage social media marketing to drive traffic to their online websites. Many articles have provided insights into the types of customers who shop online. These articles are based on research conducted on users in different countries and mostly focus on the factors that influence customers' decisions (Khalilinezhad et al., 2021).

3.4 PROBLEM DEFINITION

In 2015, more than half of the Dutch people used electronic shopping forms and achieved many benefits through this. The increasing tendency to buy goods online, more money savings, and the wide range of products have increased the number of Internet users, and this has also been very effective in increasing the number of activities being advertised in this area. The regularity of the number of activities being advertised is very effective in the development of an online store. Retaining old customers and acquiring new customers is a serious challenge for store owners. Industry publications assume that analyzing target customer groups is vital and necessary for a continuous and successful presence in the market.

A typology is a set of rules that allows grouping people, objects, and phenomena into certain types. Customer typologies can be conceptualized as a collective concept with the aim of identifying different customer groups in order to focus marketing efforts on each customer segment. These groups should be as unified and unified as possible in their own context (minimum internal deviation) and in their possible definitions with each other (maximum external deviation).

As an exploratory tool, typology should be operational and problem-oriented, so there is no universal and general recognition of types. The specific nature of typologies stems from the fact that potential and actual customer behaviors may vary depending on the product or situation, and the fundamental problem in the recognition of types is the complex number of criteria that can lead to impractical connections and consequently less effective results.

One of the objectives of this research is to develop a first-hand decision support system for online customer segmentation that extends and expands the operational capabilities of the proposed system to an acceptable level. In terms of methodology, complexity, it offers a single and unique approach that combines two classical customer segmentation methods and graphical cell automation in one system and provides more opportunities for collecting practical data.

3.4.1 Methodology

Customer adaptation and coordination are the key and important factors in sales. The rapid development of the Internet and related technologies has greatly increased the opportunities for marketing and sales science and changed the way businesses manage relationships with their customers. Understanding and recognizing a consistent approach for each user is very short and difficult. Reaching a large number of online customers is only possible by understanding their behavior, which involves all activities involved in acquiring and subsequently using and distributing products and services in an advertisement related to decisions and putting them into practice. Given the level of knowledge of customers through their profile information, relevant content can be presented and replaced on websites based on their preferences.

Various and numerous attempts have been made to model and interpret customer behavior in online shopping. For example, behavioral instability of customers was assessed by considering individual characteristics and characteristics such as perceived risk or trust. There are still gaps in our understanding of how people are involved in purchasing decisions and how these processes evolve. As a result, we see a need to change the way we examine the actual behavior of online customers and identify the limitations of customers based on their decision-making styles in online shopping research.

This study used two variables, customer knowledge and decision-making style, in developing an understanding the types of online customers. Knowledge about customer products is an individual characteristic that affects the purchasing process. However, the literature is full of contradictions. Some studies have shown that product knowledge influences online behavior, while other studies have shown no significant relationship between them. Decision-making style, which can be defined as the tendency to encounter or increase the decision, is an individual characteristic that has been shown to be a factor in decision-making. By considering these two characteristics, differences in the purchase decision process can be explained. There are studies on the differences between online customers that have led to the emergence of different types of online customers. In this work, the authors used the dimensions and quality of services and electronic services to describe customer segments based on network orientation. Most online research has used cluster analysis to identify types for market segmentation.

In addition to identifying customer types, previous customers had used general segmentation principles such as store size and height, motivation, online shopping, tool or function of implementing suggested electronic aspects, shopping focus, and modeling user behaviors. Therefore, the online typology proposed by customers was more consistent with the previous method, and most of them had used a general basis for segmentation. Because the general basis for segmentation was not based on products, services, or conditions, but on the specific segmentation basis of each product that was related to the customer, product, or condition, the latter is very useful and important for website managers.

3.4.2 Data Structure

This is a concept of data discovery and machine learning that is derived from the broader concept of inertial classification (learning without reality) and is a set of methods used to technically classify homogeneous subsets of objects from a population, which is called clustering. The goal of this cluster analysis is to divide the region into subgroups (clusters). Objects within the presented subset should be as similar as possible, but they are different from objects in other subsets. In the past, clustering was done in a subjective way based on the reader's perception and judgment. Automatic classification is a very young field that has dynamically developed as a scientific discipline. Currently, automatic classification is growing into an independent scientific discipline, as evidenced by fully developed journals such as the *Journal of Classification*. One reason for the differentiation of algorithms is the lack of a clear definition of a branch. There are different types of branches (spherical, linear, etc.). Also, different implementations use different types of data (e.g., connected and disjoint), so different methods are needed to represent objects in terms of average values or states. Typically, branching algorithms operate on one of the input structures. The first structure represents objects in terms of average values or states (such as gender, age, weight, and color). Such calculations are expressed by a branch of dimensions $m \times n$, where n represents the number of objects to be divided, and m is the number of states. The rows are associated with objects, while the columns are associated with states. The second structure is a set of distances that must be determined for each pair of objects. Branch analysis is a multidimensional method that seeks to classify objects (samples) based on the calculated state values into several similar groups, with objects belonging to the same group. They may often be useful in identifying the right treatment approach. In marketing, it is better to identify different nodes of strong core customers. Branch analysis does not allow for the distinction between case-specific and non-case-specific variables. This is why the choice of variables to include in the analysis is important, and the branches created can depend on the variables used. The choice of a particular method of branch analysis is made by the nature of the data and the expected form of the results. Branch analysis algorithms fall into several categories: staircase methods, non-staircase methods, and chaotic branch analysis. In staircase methods, for a set of samples, the staircase algorithm creates classifications called:

- Dense methods, in which each object is a separate branch. Two close (very similar) branches will remain connected until all objects are in the same group. Finally, the best number of branches is obtained from all the selected solutions. Densification processes create a similar matrix of classified objects.
- Contrastive methods – Initially, all objects belong to a single cluster. With each subsequent step, smaller subsets are created so that the objects form a distinct cluster.

3.4.2.1 K-Means Algorithm

The most common algorithm of non-stepwise methods is the K-means algorithm. In using this method in this group, it is necessary to specify the number of clusters into which the data will be distributed. Non-stepwise methods are usually used for large data sets. During the distribution of the sample, they can move between clusters, but in the stepwise method, a sample once assigned cannot be moved to another cluster. The disadvantage of non-stepwise methods is the need to specify the number of clusters, which may require a unit with several clusters.

The methods of chaotic branch analysis can assign an item to more than one cluster, which is why chaotic branching algorithms are used in classification. In this regard, chaotic branching analysis methods differ from classical branching methods in that the resulting classification is an unconnected node, which in effect causes each factor to belong to one and only one category.

3.4.2.2 Automatic Cell Diagram

The automatic cell diagram is defined as CA = (S, d, FCA), which is described as follows:

- A cellular network in d-dimensional space ($d \geq 1$)
- The automatic state depends on the set of states s_i of individual cells

- The FCA rule defines the state of cell v at time $t + 1$, which is related to the state of this cell and its neighbor $N(v - i)$ at time t
- A classical automatic cell is defined
- A known set of states for each cell
- A known set of rules (FCA) for each cell of the complete network
- An automatic network that does not change over time
- A known neighbor diagram for the entire network

If at least one of the above properties is not true, the automatic cell is not classical, or, in other words, it is non-homogeneous. The graph-based automatic cell was proposed in, where the author stated that all positions in space are in some way reflections of the entire space. The implementation of automatic cellular graphs in social networks was presented. Let us consider a weighted directed graph:

- G(V, E, K, α) is defined by a set of nodes (vectors) V, edges E, by a set of weights K, and an edge weight function α.
- The edges E of the graph G are defined by a Cartesian square, while $V_i v_j \neq v_i v_j$ says that the graph is not a directed graph.
- The function α (the weight function) defines the weights of the edges of the graph G, where K ∈ N or more generally K ∈ R.

3.4.3 Proposed System

The online store is a part of the developed system that has the following functions:

- New visitors who have visited the store for the first time – Every user who visits the store's home page for the first time is shown a short tutorial regarding the search.
- Allow customers to sign up – Any unregistered user can be authenticated after entering their username, personal email address, and password in the registration form.
- Allow customers to log in – Any registered (but not logged in) user can log in using their username and password.
- Allow customers to log off – This function enables the display of the catalog of products imported into the data packages. Each product belongs to a category and has a name and a manufacturer name, and the products are covered by the data package. The latest products are listed on the home page.
- Enables CART management – Each user adds or removes items from the cart, and in case of a non-logged-in user, the seed is stored in cookies, and the logged-in user's cart conditions are stored in the data package.
- Enables purchasing products in the cart – A logged-in user can purchase a product. Transaction confirmation is done by simply pressing a button in the cart view.
- Has an executive panel – Enables data package management (adding edited products).

- Recording visited pages and the user – The system shows user behavior by writing the URL for the pages visited by the participants.

3.5 DATA ANALYSIS AND FINDINGS

Analyzing customer behavior is one of the ways to understand the market and its potential. Therefore, using basic data to better communicate with customers and their characteristics is an important tool in customer relationship management. Therefore, using databases that contain customer data can be a significant asset. Customer segmentation means dividing the target market into subgroups, each of which can be managed with an appropriate strategy. Customer clustering will be a fundamental step in converting massive and complex data into programmable data. By categorizing based on customer similarities, costs can be reduced, and their satisfaction can be ensured. Customer segmentation with a data mining approach will increase customer value in the relevant area and at a certain level. Proper customer recognition allows companies to offer special tariffs, privileges, and facilities in accordance with the characteristics of unusual and special segmented customers.

Cluster analysis or clustering is one of the most basic branches of unsupervised learning in the field of mining. Cluster analysis is used in various fields, one of the most important of which is market research and marketing studies. Clustering is a type of indirect data mining operation. In most data mining methods, it starts with a training set, with the help of which we try to create a model to segment the data and then predict new data; but in clustering, there are no pre-existing categories, and, in fact, the variables are not divided into two independent and dependent classes. In clustering, the focus is on groups of data that are similar to each other, so that by discovering these similarities, behaviors can be better identified, and based on this knowledge, a better decision can be made. Various methods with different approaches have been developed for performing cluster analysis of data, which we will discuss below.

3.5.1 Implementation

The prepared prototype of the website software, in the form of an online store, makes it possible to collect background data of customers in a way that is surprising to the person who participated in the experiment. When this store is used, the pages visited by the customers are recorded. It is assumed that the duration of the test is equal to the duration of the website session. A website session is a record that is kept on the server for a period of time, including information about the connection with the website browser. When the browser is opened, a session is created, and a unit number is determined, and when the browser is closed, the session expires. This mechanism is created by the session number on the server. A text file is created in the directory. Each change in the displayed page is recorded in the mentioned file, resulting in the creation of a directory tree in which a directory belongs to a participant, which is used in the branching.

TABLE 3.1
User Activity Log

Activity Type	Keyword
Show Promotional Products	Advertisements
Show New Products	New
Show Gallery Product Images	Gallery
Use Search, Filter, and Sort to Arrange	Products
Show Product List on a Specified Page	Specified Category
Add to Cart	Add
Remove from Cart	Remove
Click on Preview Invoice	Sales Invoice
Login/Sign Up	Login
Click on Manual Purchase Link	How to
View User Profile	Account
Complete Order Purchase Process	Order

3.5.2 User Activity Recording

In order to identify the types of customers, the actions performed by the users of the online store should be recorded. The following activities are recorded: visiting the subpage containing the product being advertised, visiting the product details page to view the product gallery, using search/sort/filter, viewing the product list page by category, adding an item to the cart, clicking on the order summary (view cart), logging in/signing up, clicking on the store manual, viewing the user profile, and completing a transaction. Each of these operations is associated with a keyword that is written in the user file (Table 3.1).

For the category label feature, users are categorized based on whether they have not yet created a cart (first category), created a cart but have not yet purchased (second category), and created a cart and finalized their purchase (third category).

3.5.3 Clustering Using K-Means Algorithm

The clustering was performed using the k-means tool, MATLAB. The function (k-mean) is used to divide the algorithm (k-mean). The input to this function is the data matrix and the number of branches into which the data is divided. The recorded data should be written in the form of a matrix. The number of rows in this matrix corresponds to the number of participants in the estimation. The number of columns corresponds to the number of user activities (Table 3.2). Each column represents the value of one of the states. The first column is the urls patterns vector, which contains the keywords related to the recorded activity. Then the system calculates the occurrences of each of the defined keywords in the content. This number is entered in the data matrix, and each row is for a file, and each column is for a single number, which determines how many times the keyword (state) occurred. Finally, each row of the data matrix is divided by its total. As a result, for each instance, the percentage of occurrences of individual activities relative to all activities performed by the customer is defined. Such a matrix is expressed in the form of a parameter for the (k-mean) function.

3.5.4 Prediction

In order to more effectively coordinate and adapt the displayed products to the needs of the customer browsing the contents of the online store, a prediction mechanism is proposed, in which a group of active customers can be determined. This mechanism consists of a continuous and integrated writing of user activities to previously visited graph structures stored in the data package. Based on this adaptation, the current customer is pre-classified, and active products are displayed in an acceptable form for the customer group.

3.5.5 Classification

Another system agent is the classifier, which creates lists and retrieves data that is designed for the matrix. Each user is associated with a row in this matrix. The data forms matrices that are sent to the network function. This function returns groups that identify the types of customers. This function is based on the branch analysis method.

In addition, the automatic cell diagram is used here to predict which groups of users a newly arrived customer may belong to. The classification mechanism, called segmentation or customer grouping, is based on the overall activity of customers in the online store. The final result is another sample that is stored in the data set and used to predict when analyzing the behavior of other customers in the online store.

3.5.6 Experimental Results

The aim of this study is to identify the types of customers in online stores based on the sites they have visited. The result of this study is the characteristics of groups of customers. This study was conducted under production conditions (non-laboratory) to predict the natural environment during the experiment. The researcher does not know the details of the experiment objective and only has the title of the experiment and the link to the website page from which he started his tests. He does not have information about the movements and the method of supervision and inspection. When the execution operation for customer classification is started, a short instruction is displayed in which the customer is asked to perform the experiment seamlessly (without interruption) and only close the browser after the tests are completed. The customer is asked to perform the test only once. This study is based on the understanding and inference of the concept of a framework of online shopping (Figure 3.1). This study used a randomly selected group with an average age between 20 and 40 years. These individuals have IT knowledge and are authorized to establish an online store. As a result of this research, data in the form of keywords related to the actions performed by each participant were collected. In order to determine the best number of branches, the effects of dividing the data into 3-4-5-6-7 groups were analyzed. To find out how

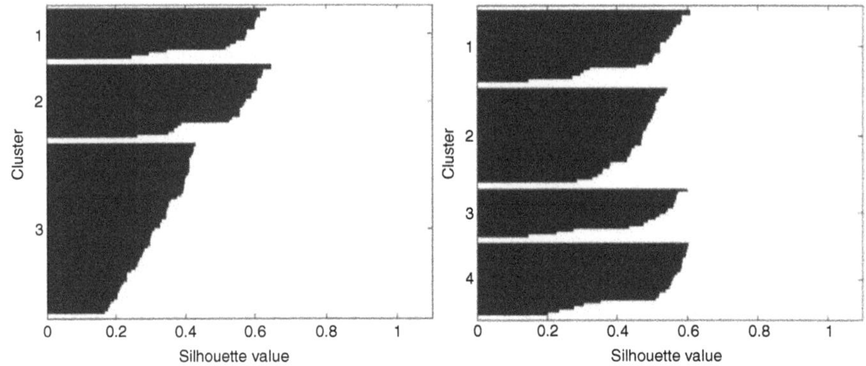

FIGURE 3.1 Dividing the data into three and four groups.

far the branches were separated, the silhouette method was used. Figures 3.2 and 3.3 show a calculation of the degree of proximity to the neighboring branch in the sample. This calculation has values from 1 to −1. A value less than 0 indicates that the sample was assigned to the wrong branch.

Average matrix values obtained for the number of different groups are shown in Table 3.2.

If the calculated value is higher, the sample will find better compatibility.

Based on the above figures, it was concluded that the best unit was divided into four groups. For the four categories, all points were correctly distributed (the calculated value reached a value below 0). To accurately determine the best value of the groups, the average value for each unit was calculated (Table 3.3). The highest value was obtained by dividing into five categories. However, in the case of this system, the points reached negative values. The next division with the highest calculation is the division into four categories. So, in the subsequent analysis, this division was calculated. During the research test, a campaign was formed on the website. The activities that had the greatest impact on the groups were: showing promotional products, displaying new products, viewing galleries, and ordering products.

These four activities accounted for 95.9% of the variance. Other activities were also considered for the final classification of customers. The results of this study are organized in Table 3.3.

The purpose of customer segmentation in this thesis is to develop an initial list of decision support systems for customer segmentation. Accordingly, the methodological assumptions of the conceptual framework were proposed. The combination of customer segmentation methods and an automated cell diagram in a decision support system was demonstrated in a large-scale practical implementation. A pilot study was conducted. The results showed that in a logical division, there are four groups: opportunity hunters, new product followers, customers who are eager to visit online stores, and information hunters.

It is worth noting that the generated design principles of the IT system, apart from the methodological layer shown, have great practical possibilities. The conceptual framework, along with a set of operational assumptions, provides complete practical

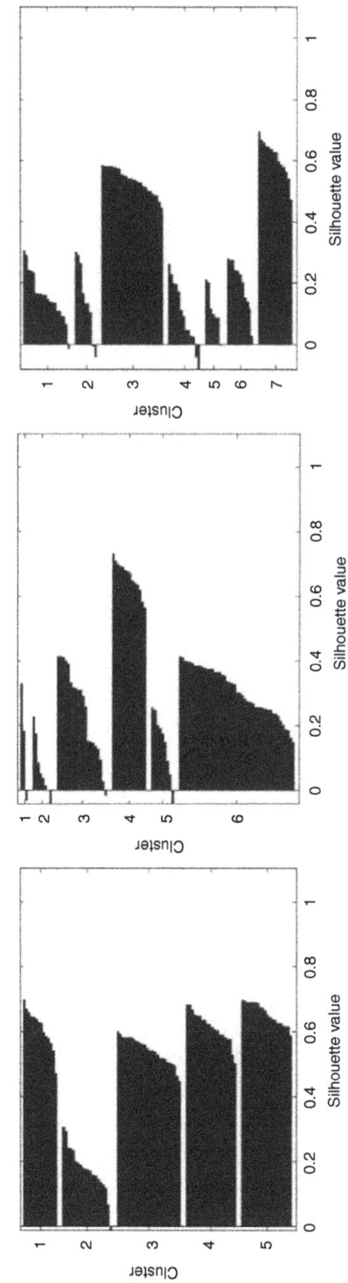

FIGURE 3.2 Dividing the data into five, six, and seven groups.

TABLE 3.2
Average Matrix Values Obtained for the Number of Different Groups

The Number of Clusters	The Value of Measure
3	0.4086
4	0.4808
5	0.5138
6	0.3062
7	0.3111

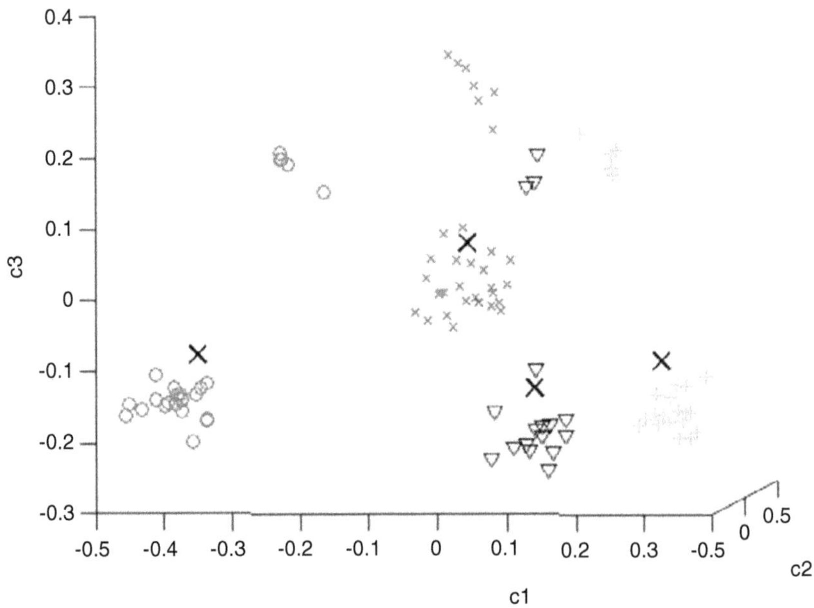

FIGURE 3.3 Observation of branches and their dispersion.

solutions to meet operational needs and forms a formal and independent platform for the customer segmentation system, regardless of the size of the customer group, as well as the profile and breadth of the product in the Entrotti store. At the beginning of this research, possibilities for the growth and improvement of the solutions were presented and examined. It seems that it is very interesting to compare other master segmentation methods with the automatic cell diagram. They can also provide the most accurate assessment. The decision support system can also be introduced based on the technological and methodological basis of the presented framework and its application in a real e-commerce environment.

The purpose of categorizing customers in this study was to determine their type of motivation to purchase and to provide advertising space and web design that are

TABLE 3.3
Initial Results of Customer Segmentation

	Categories	Characteristics of Categories
1	Opportunity hunters	This group includes people who try to buy products at low prices. Such customers sacrifice quality for a lower price.
2	New product followers	Users in this group try to choose manufacturers' offers.
3	Customers who are eager to visit online stores	People who spend their free time browsing the Internet and pay special attention to the product galleries provided.
4	Information hunters	This group includes people who compare products, prices, and product characteristics before buying.

TABLE 3.4
Determining the Content Marketing Strategy Based on the Characteristics of the Categories

Customer Type	Dedicated Advertising Strategy
Opportunity hunters	Special discounts – 2nd- and 3rd-grade products – daily offers according to the change in currency prices and market inventory
New product seekers	Advertising the latest products and goods produced in the world – business news – products with multi-purpose use
Customers who are eager to visit online stores	Offering installment purchases – the latest products and services – using the latest information and news from the media world in content design – special offers
Information hunters	Practical explanations of the product information area – need assessment and definition of requirements to make the product efficient – compilation of new research information related to the product

appropriate to their behavioral preferences. In this way, the design and compilation of content for the website are tailored to the number of groups specified in the clustering. Experienced users initially enter a home page whose front-end design provides a general reference to the nature of the desired brand and the type of service it provides, after which they can enter their own section by logging in. Also, for new users, a general design that includes examples of the different segments can be displayed.

A lot of research has been done in the field of customer typology, and it generally ended at this stage. However, in the present study, after typology of customers and website visitors, we consider creating a unique space for the identified categories or clusters so that the user is constantly in their own advertising space, according to their type of interest in purchasing.

This type of dedicated content design for the user can encourage them to buy. We see this space creation by famous and reputable global brands on websites and social networks.

Based on the identification of customers categorized in this study, the type of content designed for each category should have the following characteristics, according to Table 3.4.

3.6 CONCLUSIONS

Consumer behavior is one of the most popular trends in psychology and has attracted the attention of many experts, manufacturers, and service institutions in recent years. Some of the most important and main areas of this field are: expectations, tastes, and needs of consumers (customers). With the help of this science, it is possible to examine the factors affecting the choice of the Calara brand by identifying consumer behavior, as well as analyzing the cultural and social factors affecting consumer decision-making. By correctly applying methods for studying consumer behavior and predicting behaviors related to purchasing, it is possible to create and present better products, which ultimately leads to increased sales and customer satisfaction. Familiarity with consumer behavior affects the quality and quantity of an organization's work.

Making the right decisions and creating effective marketing and sales strategies and tactics depend on organizations correctly identifying and understanding consumers and buyers.

Today, marketing researchers are constantly trying to identify consumer behavior and, accordingly, adopt appropriate solutions for better and more effective sales and increase market share. In order to meet demand and maintain and expand market share, manufacturers must be aware of the perceptual, rational, and behavioral layers of potential buyers so that they can produce, distribute, and promote products in accordance with their decision-making structures and mechanisms. Product innovators and designers, simply by recognizing consumer behavioral patterns, can become aware of the angles of expectations and satisfaction of individuals and make their use smoother by modeling. One way to better understand consumers is to use a systems approach to explain the behavior of individuals and organizations. The consumer is a complex system of motivations, tendencies, behaviors, and performance with a wide range of responses and reactions. The consumer is influenced by the connection with larger social, economic, cultural, and political systems.

In the present study, we aimed to identify the behavior of online customers, their decision-making styles, and their segmentation into meaningful groups so that we could provide them with the best content in online web stores based on their shopping needs. Our goal was to improve the knowledge of buying and selling in the context of e-commerce, so that, given the stagnation of traditional markets in Iran and the decrease in the volume of transactions, by launching a smart part of e-commerce, we could focus transactions on online platforms so that with these methods we could have maximum savings on costs and fuel and energy consumption, and we could advertise, offer, and deliver products to customers through different channels.

3.6.1 Future Suggestions

1. Various tools, such as categorization and decision support systems in neural networks, can be used to optimize and classify customers.
2. The statistical population should be selected from local websites that have a wider range than other stores, so that the methods studied are more efficient.
3. A blend of electronic marketing called content marketing can be used to determine advertising strategies and content production.

4 Customer Churn in Social Media Using Integrated Association Rules and Neural Networks

4.1 INTRODUCTION

In today's competitive market, customer retention will play a significant role in the growth and development of organizations. Some customers leave the system for known or unknown reasons, reducing the system's profit and survival. Predicting when customers will turn away and developing strategies can play a significant role in customer retention. Some customers leave an organization for various reasons and refer to another organization to meet their needs. Attracting new customers is much more expensive for any organization than keeping current customers. If the reasons for customer turnover can be identified, strategies and programs can be developed for the survival of such customers.

The most valuable asset of a company is its customers. As a result, customer relationship management (CRM) is an important task that can help identify the causes of customer turnover. CRM is a strategic approach to customer relationship management that aims to create profitable and long-term relationships through the use of information technology. The goal of CRM is to retain customers by identifying and understanding the valuable segments of the customer base. After identifying the causes of customer churn, appropriate marketing strategies can be applied to increase customer satisfaction and maintain customer loyalty.

Today, online shopping in the world has become more economical and convenient than in the past because, in addition to ease of shopping and saving time, it also allows consumers to compare prices and quality, and thus customer satisfaction has increased, and they often prefer to make their purchases via the Internet. In many cases, even buyers choose their products in stores and then purchase them online because it will be more economical for them. In some parts of the world, such as the United States, online sales have surpassed face-to-face sales, and this has led suppliers to offer their products online in addition to physical sales.

In today's world, traditional advertising has given way to modern and new advertising, so if a company or store wants to compete and surpass its competitors, it considers it necessary to be active in this type of advertising. One of the most important factors in customer rejection from an online store is negative advertising about that store on social networks. These social networks include Instagram and

DOI: 10.1201/9781003740148-4

Telegram. In reputable and popular social networks, the speed of news and information transmission is very high, so negative advertising can greatly affect customer churn and rejection.

The importance of advertising and marketing has increased more than ever due to the growth of production and competition in the market. Given the rapid expansion of social networks such as Instagram and the opening of the information space, mass media can no longer be used as the only news and information channels for audiences. With the emergence of social media, the tools and methods for communicating with customers have changed greatly. Therefore, companies should learn how to use social media in their business plan.

This study provides a customer relationship management system to examine customer churn in the face of social media advertising for an online business, which is a digital goods online store. It also uses an association rules model to analyze inputs and a neural network to model and obtain output behavior. Then, the result of predicting customer churn is analyzed to determine the level of loyalty of the store's customers and the causes of customer churn.

Our focus in this study is on digital goods online store advertisements on social networks such as Instagram. Finally, we provide suggestions based on the results to improve the store's business and thereby increase revenue and sales.

Today, the emergence of online stores that offer various goods in the virtual world is increasing, so that the number of online stores is now more than 900. For online stores to continue operating in this competitive business environment, business processes must be carried out in line with customer expectations, which creates customer satisfaction. In the modern economy, the priority of business activities is customer relationship management, which is considered a business strategy. In this thesis, the purpose of clustering is to group a large set of customers according to their purchase history from the online store Digikala. Then, this model is used to calculate customer behavior under the influence of social media advertising with the help of association rules. Finally, the results are used to improve processes to prevent customer churn.

In this study, we intend to examine the advertisements of the Digikala online store on social networks in order to obtain the extent of the effects of these advertisements on customer loyalty and churn. For this, we will first examine the advertisements on the Instagram social network. For this, we will use the extraction of people's opinions from comments and examine the amount of product sales in a specific time period, and then we will analyze the data using the K-means clustering algorithm with association rules. After that, we test the input data with the integrated neural network and analyze and review the results. Finally, we analyze these results using sales techniques and propose a solution for more sales and customer retention.

4.1.1 Customer Churn

Customer churn or churn, which is also referred to under titles such as customer attrition and customer abandonment, is a business term used for customer loss. In fact, customer churn is a situation where customers of an organization decide to leave that organization.

Churn is considered equivalent to the word Churn in English, which is derived from the two words Change, meaning change, and Turn, meaning rotation, and refers to the fact that a customer changes their service provider by turning away from the current service provider. However, a more precise definition of this term in the business environment is considered equivalent to a customer changing service providers or a customer's tendency to sever ties with an organization over a period of time.

4.1.2 Association Rules

Associative analysis (dependence) is a data mining algorithm that can be used to discover the dependencies between one or more features and thus quantify the relationships between them.

4.1.3 Neural Networks

An idea for processing information inspired by the biological nervous system, a neural network processes information in a way similar to the brain. The key element of this idea is the new structure of the information processing system. This system consists of a large number of highly interconnected processing elements called neurons that work together to solve a problem. Artificial neural networks, like humans, learn by example, and a neural network is adjusted during a learning process to perform specific tasks such as pattern recognition and information classification. In biological systems, learning is accompanied by adjustments in the synaptic connections between neurons. This method is also used in neural networks.

4.2 BACKGROUND

Since the popularity and use of social networks is currently increasing, companies are using them to communicate with customers, manage their brands, increase loyalty, and customer satisfaction. Companies can also use social networks to obtain their customers' opinions, desires, interests, tastes, and preferences. "Samira Rezaei and colleagues used text mining as a method to analyze customer opinions and interests in social networks, whose data was related to the Iranian digital products market" (Akbari et al., 2013).

In this study, two separate clustering techniques and association rules were used to analyze Iranian customers' opinions about mobile phones from six different brands (Samsung, LG, Apple, HTC, Sony, and Huawei). The results showed that from the customers' perspective, no mobile brand is superior, and each of them has positive and negative features. However, Iranian customers have positive feedback on features such as design quality, screen, price, sound, image, and camera, as well as negative feedback on the feature of battery life. Association rules were also used to extract hidden relationships that existed among all mobile phones. The results of extracting association rules for mobile phones provided a strategic guide for digital product service providers to incorporate their customers' opinions into their products and increase customer satisfaction. These rules are a buying guide. Customers were also able to purchase a phone with their desired features and advantages, and on the other hand, customers could have a general understanding of the mobile phones that are offered to the market and the products that will be introduced to the market, and they could easily recognize the advantages and disadvantages of each.

In a study titled "A Model for Measuring the Effect of Bank Brand Trustworthiness on Customer Loyalty Commitment", they show that brand trustworthiness plays a very important and key role in improving customers' behavioral tendencies. This means that increasing customer satisfaction causes them to advertise and recommend their bank to others and also reduces their tendency to change banks. Also, increasing customer loyalty reduces their tendency to change banks. Therefore, managers should pay special attention to brand trustworthiness and consider its important role in customer relationship management and design more appropriate strategies and programs in order to develop their loyal customers (Fazlollahtabar, 2012a). In a study titled "Evaluating Brand Personality on Customers' Attitudinal and Behavioral Loyalty to the Hyperstar Brand", they concluded that out of the five dimensions of brand personality (responsibility, dynamism, emotionality, boldness, and simplicity), four dimensions of responsibility, dynamism, emotion, and boldness have an effect on attitudinal loyalty and attitudinal loyalty to the brand, but brand simplicity has an effect on attitudinal loyalty has no effect. Meanwhile, among the dimensions of brand personality, only dynamism has an effect on behavioral loyalty.

A study examined the effect of customer satisfaction and organizational image on customer loyalty in restaurants in Abhar city, and it was finally determined that there is a significant relationship between customer satisfaction and organizational image with customer loyalty. With the improvement of living standards and the change of people's sustainable living patterns, hairdressing services are now widely used by people.

In this study, 11 clusters were formed with the SOM technique and then used to apply the K-means technique to the RFM model. The results of the RFM model analysis showed that hair salon customers can be divided into four types of customers, including loyal customers, potential customers, new customers, and lost customers, and unique marketing strategies were created for these four types of customers (Mehrmanesh et al., 2013).

Liu has tried to overcome this deficiency to some extent in his research to identify the reasons for customer churn in a Taiwanese airline. He initially extracted important factors regarding customer behavior in the airline market through survey research and then tried to identify the most important factors in this area with the help of the factor analysis method. Finally, based on these factors, a main questionnaire was designed that collects the desired data in three main categories: demographic data, customer satisfaction components, and their decision to purchase in the future. And finally, an attempt was made to prepare a set of rules in the form of if, then, by applying a classification technique to the results of the questionnaire that would help to better understand the phenomenon of customer churn from the organization and the reasons for its occurrence.

A study has been conducted to analyze how social media supports the development of customer relationship management. The important role of social media, the use, and the emergence of social networks have been examined. Based on case studies and a questionnaire, this chapter has examined the necessity of transitioning from traditional customer relationship management to social customer relationship management.

Social media use for CRM and business performance satisfaction: The moderating role of social skills and social media sales intensity Examining the relationship between the intensity of social media use for customer relationship management and business performance satisfaction is investigated.

4.3 RESEARCH MODEL

According to the materials of the previous two sections, in the first section, we explained the generalities of the research, and in the second section, we talked about the literature and background of the research, which led us to a problem. Our research problem includes details that we are going to explain in this section. Finally, we will describe the tools used in the research.

With the advancement of technology, people are looking for the most convenient ways to do their daily tasks, including online shopping. There are many reasons to shop online instead of going to stores. Online shopping is more convenient, meaning that it is always available and possible 24 hours a day, 7 days a week, and in any weather conditions without having to look for a parking space and endure the traffic jams of these days. Other reasons include the ease of searching for desired products, having a variety of choices from different brands, and the ability to compare products and their prices with each other and reach a good decision to buy. Other benefits of online shopping include saving time, energy, and fuel. As a result, online shopping has become the first choice of most people, and most consumers prefer online shopping, where they can buy what they need with a few simple clicks, and all they need to do is wait for it to be delivered to the desired address, which may even be a long distance from them (Fazlollahtabar and Mohseninejad, 2015).

So, considering the advantages of online shopping, we can conclude that the competition between online stores is very high, and customers play a key role in this business. In fact, the online selection of products is up to consumers and customers; they have the power of choice and the world of online retailers at their fingertips. In the meantime, if the online store does not meet their needs, customers will quickly turn to a competitor's store. This highlights the major issue of customer loyalty, which is also the issue we are discussing in this research – increasing loyalty in online businesses. Increasing customer loyalty plays an important role in the success of the organization, especially when customer purchases are not equivalent to long-term success. The concepts of satisfaction and loyalty to a business website are the main concerns of marketers. E-loyalty is the customer's favorable attitude toward the online store website, which forces the customer to repeat the purchase behavior. A loyal customer always intends to buy from a specific website and will not be transferred to another website, which is a great competitive advantage. Keeping an existing customer is cheaper and more economical than searching for a new one. Loyal customers are more likely to spread favorable messages about positive past service experiences and recommend the company to non-loyal and new customers, creating the possibility of increased advertising at no additional cost to the service provider. Therefore, a loyal customer is considered a major asset for online businesses. We conclude that e-loyalty is a strategic necessity for long-term business success, and, in general, customer retention is directly related to the profitability of online retailers, so investing in this area will be effective and beneficial.

This is where the use of a customer relationship management system becomes necessary to align business processes with customer expectations and needs and to run the organization around the customer. Retailers must be responsive to customer needs and manage and anticipate them. In this study, we analyze the opinions, reviews, and interests of customers and consumers who have shared their experiences of purchasing products online on the store's Instagram social network to measure customer loyalty. The level of interest and opinions of people about products on the social network Instagram is an important source of information for customer decision-making. According to the research conducted, it has been found that about 90% of users read reviews, and more than 80% of them trust the opinions of consumers who are considered strangers, which is surprising. In fact, users are interested in these reviews and opinions because they provide a third-party insight into the quality and experience of their product. So these reviews and opinions are very powerful, and when used effectively, they can help the store increase sales. It is clear that ignoring these opinions by the organization as a marketing opportunity is considered to be ignoring a large percentage of their buyers. Reviews and opinions can eliminate any doubts that customers have about a product or the choice of that product. Ultimately, it can be said that customers play a very important role in online business, and paying attention to their opinions and reviews is also considered extremely important.

Making sense of, collecting, maintaining, and organizing data plays an important role in organizations, where most of our data consists of customer reviews and comments, which are unstructured texts. In this study, we use data mining software to extract user opinions, which we will discuss in the next steps. Next, using these techniques, we will examine consumer opinions and carry out processes with the goals of examining user opinions and suggestions, as well as analyzing the strengths and weaknesses of products. Then, we cluster the data using the K-means algorithm and with the help of association rules. Finally, we examine the clustered data through an artificial neural network and analyze the results to improve this business and prevent customer churn, and we will provide useful suggestions.

4.3.1 Modeling

In the process of purchasing from online stores, the user searches and examines different sections of a website to find the product they want. In this process, the user will be faced with various and diverse options and may easily pass over a number of high-quality products due to the lack of comments. In this research, finding the desired product for analysis is considered to be one of the initial and important steps. In this stage, best-selling, most visited, and most popular products that also have many comments have been used. After reviewing and selecting the desired data, we move on to the next stage, which is extracting and translating them.

Web mining is the application of data mining techniques to identify patterns on the web. Web mining is divided into three categories: web content mining, web application mining, and web structure mining. Web content mining is the process of extracting useful information from the content of web documents. The content of a web document corresponds to the concepts that the document seeks to convey to users. This content can include text, images, videos, audio, or structured records such as lists and tables. Among them, text mining has been researched more than any other field.

In this study, we use the web content mining method to extract comments and take the help of the Web Scraping technique (explained in detail below) using the Data Miner extension in Google Chrome. Then, we examine and analyze the extracted inputs with the help of association rules. For this, we use the RapidMiner software and the FP-Growth algorithm. In the next step, clustering is performed using the K-means algorithm and through the RapidMiner software.

The clustered data is entered into the MATLAB software to analyze and examine them. In this study, the Multi-Layer Perceptron (MLP) artificial neural network has been used for this test.

Finally, by analyzing the output from this neural network, the prediction of attrition is performed. We will explain these processes in the following.

4.3.2 Web Scraping

Web Scraping is a technique for extracting large amounts of data from websites, where the extracted data is saved in a local file on your computer or in a spreadsheet format. This technique is also known as Screen Scraping, Web Data Extraction, and Web Harvesting. The data displayed by most websites can only be viewed using a web browser. They do not provide a function to save a copy of this data for personal use. The only option is to manually copy and paste the data, which is a very tedious task and may take several hours or sometimes several days to complete.

Web Scraping is actually an automation technique of this process, so that instead of manually copying data from websites, Web Scraping tools do the same thing in a matter of seconds. Web Scraping tools automatically load and extract data from multiple website pages based on your needs. It is either custom-built for a specific website or can be configured to work with any website. With the click of a button, you can easily transfer the information from the website to a file on your computer. There are many software tools that can be used to customize Web Scraping solutions. The extracted data is saved as tables in Excel. Then we need to clean and prepare this saved data using Excel. With some initial preprocessing, we prepare the data for entering into the software.

4.3.3 Association Rules

In data mining, and in particular association rule mining, various constraints (such as thresholds) are applied to various metrics related to importance and interestingness in order to select interesting rules from a set of possible rules. The most well-known constraints in association rule mining are the minimum threshold for support and confidence. Other metrics, such as lift and conviction, are also used in this regard.

4.3.3.1 Support

Support is an indicator of how many times an item set appears in a data set. The support X, with respect to T, is defined as the fraction of transactions t in the data set that contain the item set X.

$$\text{supp}(X) = \frac{|\{t \in T; X \subseteq t\}|}{|T|}$$

4.3.3.2 Confidence

Confidence is an indicator of how many times a rule was true. The confidence value of a rule (X→YX→Y), given a transaction set T, is the fraction of transactions containing X that also contain Y. Confidence is defined as follows.

$$\text{conf}(X \Rightarrow Y) = \text{supp}(X \cup Y)/\text{supp}(X)$$

The concept is a bit confusing because it is natural to think of probabilities of events, not of sets of items. We can rewrite sup(X∪Y) as the probability P(EX∩EY), where EX and EY are the events in which the transaction contains the sets of items X and Y, respectively. Confidence can be interpreted as an estimate of the conditional probability P(EX|EY), which is the probability of finding the RHS of the rule in transactions under this condition, which also contain the LHS.

4.3.3.3 Lift

The lift of a rule is defined as follows:

$$\text{lift}(X \Rightarrow Y) = \frac{\text{supp}(X \cup Y)}{\text{supp}(X) \times \text{supp}(Y)}$$

If the lift is greater than one, it allows the Data Miner to know the degree to which two events are dependent on each other and potentially use these rules to predict the outcome in a future data set. If the lift is less than one, it allows the Data Miner to know that the items are substitutes for each other. This means that the presence of one item has a negative effect on the presence of another item and vice versa.

The value of lift comes from the fact that it captures both the confidence of a rule and the generality of the data set. To further explain the issue, in data mining and dependency rule learning, lift is a measure that measures the effectiveness of the target model (dependency rule) in predicting or classifying items in a way that has an improved response (relative to the total population), as measured by a random selection of the target model.

4.3.3.4 Conviction

The conviction of a rule is defined as follows.

$$\text{conv}(X \rightarrow Y) = 1 - \text{supp}(Y)/1 - \text{conf}(X \rightarrow Y)$$

In this research, we intend to analyze the extracted data, which includes customer comments on the social network Instagram of the online store Digikala, through the RapidMiner software using association rules to determine the importance of the data and remove useless data, and then use the normalized data to continue the research process.

The RapidMiner software uses the FP-Growth algorithm to analyze data using association rules, which is explained below.

4.3.4 FP-Growth Algorithm

The FP-Growth algorithm uses a tree structure called an FP tree to store information about a set of alternating items. Each path from the root of this tree to one of its leaves represents a transaction in the transaction database. Items are organized in the tree in descending order of their number of occurrences in the transaction database, with items with the highest number of occurrences being placed closer to the root. The FP tree has a header table that contains all alternating items in the transaction database (along with the number of occurrences of each). This table is denoted by Header (T). Each record in the header table is the starting point of a list that links all FP-tree nodes that contain the item in this header table record.

4.3.4.1 Clustering

Clustering is one of the most important data discovery tools used in random discoveries. Currently, knowledge extraction is a major bottleneck in the knowledge engineering process. Machine learning and data mining algorithms are proposed as a way to solve this problem with the aim of extracting knowledge from data. A common approach in this field is the clustering method, which can assign symbolic decisions to new examples using existing examples for decision-making or classification. Clustering methods have gained popularity due to their inherent comprehension capabilities. The presence of comprehension capabilities is important in various ways. Understanding the domain, understanding the classification capabilities, justifying the decision, and finally the presence of symbolic rules that can be extracted from the clusters and then used in a rule-based decision-making system.

Clustering is actually an unsupervised operation. This operation is used when we are looking to find groups of similar data without having a prior prediction about the existing similarities. Clustering is usually used when we are looking to find groups of customers that are not previously known. For example, we can search for similarities in customer usage in order to group customers and identify a new service.

Clustering for most data mining methods, such as decision trees and neural networks, starts with a training set. With the help of this set, an attempt is made to create a model for segmenting the data. Then, that model is used to predict new data. The Cummins K-means method is one of the data clustering methods in data mining. Despite its simplicity, this method is considered a basic method for many other clustering methods (such as fuzzy clustering). This method is considered a proprietary and flat method. Various forms have been described for this algorithm. But they all have an iterative procedure that, for a fixed number of clusters, tries to estimate the following:

- Obtaining points as cluster centers. These points are actually the average of the points belonging to each cluster.
- Assigning each data sample to a cluster that has the shortest distance to the center of that cluster.

In a simple type of this method, first, points are randomly selected for the required number of clusters. Then, in the data, they are assigned to one of these clusters according to the degree of proximity (similarity), and thus new clusters are obtained. By repeating this procedure, new centers can be calculated for the data by averaging them in each iteration, and the data can be reassigned to new clusters. This process continues until there is no more change in the data. The following function is considered the objective function.

In the K-means algorithm, first k members (where k is the number of clusters) are randomly selected from n members as the centers of the clusters. Then the remaining n-k members are assigned to the nearest cluster. After all members are assigned, the cluster centers are recalculated and assigned to the clusters according to the new centers, and this process continues until the centers of the clusters remain constant.

4.3.5 Artificial Neural Network

In the past few decades, when computers have enabled the implementation of computational algorithms, many research works have been started by computer science experts, engineers, and mathematicians in order to simulate the computational behavior of the human brain. The results of their work are classified in a branch of artificial intelligence science and in a sub-branch of computational intelligence under the subject of "Artificial Neural Networks" (ANNs). In the subject of Artificial Neural Networks, numerous mathematical and software models inspired by the human brain have been proposed, which are used to solve a wide range of scientific, engineering, and applied problems in different fields.

In this research, we intend to use this neural network for processing and analyzing clustered data, considering the characteristics of the MLP algorithm. After analyzing the results of this algorithm, we will provide positive solutions for increasing customer loyalty and reducing their churn. We will provide a complete description of the selective neural network below.

4.3.5.1 Multi-Layer Perceptron Neural Network

One of the most basic neural models available is the Multi-Layer Perceptron (MLP) model, which simulates the transfer function of the human brain. In this type of neural network, most of the network behavior of the human brain and signal propagation have been considered, and hence, they are sometimes called Feedforward Networks. Each of the nerve cells of the human brain, called Neuron, after receiving input (from another nerve or non-neuronal cell), performs a process on it and transmits the result to another cell (neuronal or non-neuronal). This behavior continues until a specific result is achieved, which will probably eventually lead to a decision, processing, thought, or movement.

4.3.6 Research Tools

This section introduces the tools and software used in this research.

4.3.6.1 Data Miner Extension for Google Chrome Browser

Data Miner is a Google Chrome extension that helps us save information from web pages in a CSV file or Excel spreadsheet. This tool can be downloaded from https://data-miner.io or from the Google Chrome browser extensions, as shown in Figure 4.1.

Data Miner has more than 40,000 general instructions for many popular sites. Using these general instructions, you can get the structured data you need with just one click and save time, as shown in Figure 4.2. Some tools only work on specific websites, but Data Miner can be run on any web page and has a full team of experts ready to help extract information with custom instructions.
Key features:

- Extract tables and lists: Easily extract information from lists and tables on websites.
- One-click scraping: Extract data in seconds using ready-made instructions, and it can be downloaded as a CSV or xlsx file.
- Extract results from multiple pages: Automatically crawl pages and extract information from all pages.
- Scrape and open a list of URLs: Provide Data Miner with a set of URLs to scrape data from.
- Run custom JavaScript: Use custom JavaScript to clean up scraped data, such as extracting emails.
- Autofill forms: Automatically fills forms with the information it receives.

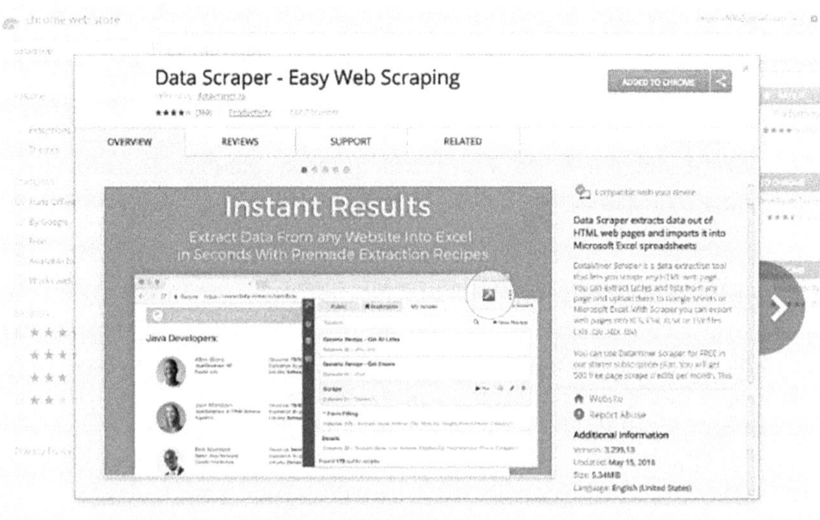

FIGURE 4.1 Data Miner extension.

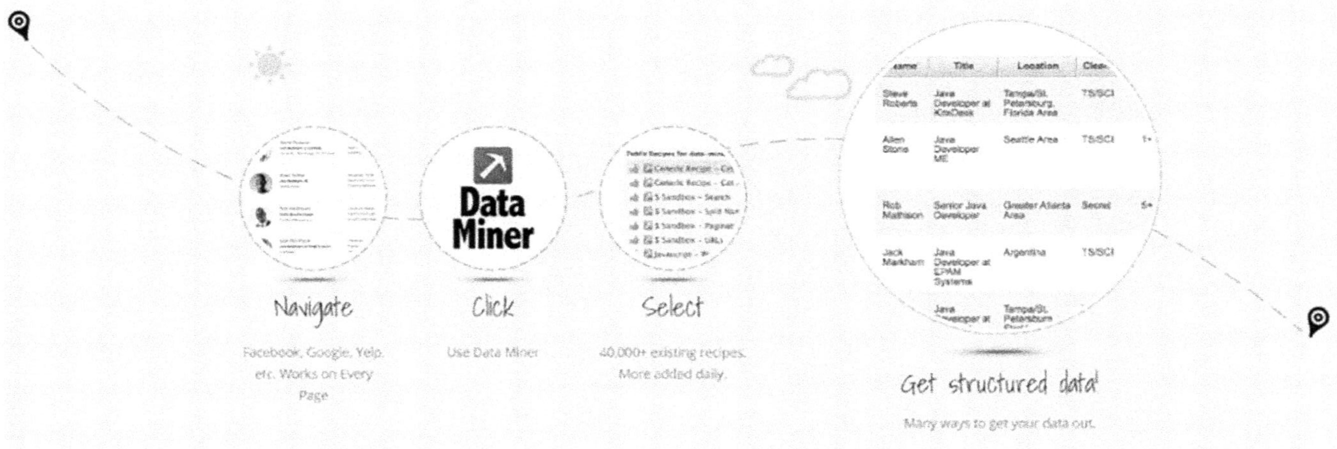

FIGURE 4.2 Data Miner process.

- Downloading full web pages: Automatically downloading the full HTML of web pages from different sites. In fact, Data Miner scrapes URLs, extracts page information, and downloads the page to your computer, downloading text, images, and every element of the page.
- Download images: Scrapes image links, extracts the image, and then sends the images to the computer. Once the images are saved, they can be uploaded to your site, as shown in the figure.

4.3.6.2 Google Translate Extension for Google Chrome Browser

Google Translate is a free multilingual translation service provided by Google for text translation. The service is available in various formats, including website, Android, and IOS mobile applications, and an API that helps developers create extensions for browsers and software applications. Figure 4.3 shows a screenshot of the Google Translate application on Android. Google Translate supports more than 100 languages at different levels and serves 500 million people daily as of May 2017.

Launched in April 2006 as an American statistical translation service, Google Translate uses reports from the United Nations and the European Union Parliament to collect linguistic data. Rather than translating directly, it first translates the text into English and then into the target language. During a translation, it looks for patterns across millions of documents to help decide on the best translation. Its accuracy has been criticized and ridiculed several times. In November 2016, Google announced that it was transforming Google Translate into a neural machine translation engine – Google Neural Machine Translation (GNMT) – that translates entire sentences at a time, not just fragments. Using this broader context helps identify the most appropriate translation and then rearranges and adjusts it to sound more like human speech with proper grammar. GNMT was initially enabled for only a few languages in 2016 and is gradually being used for other languages.

Google Translate can translate various forms of text and media, including text, speech, images, websites, or real-time video from one language to another. For some languages, Google Translate can pronounce the translated text, highlight different words and phrases in the source and target text, and act as a simple dictionary for single-word input. If the "Detect language" option is selected (the language of the text is not specified), text in an unknown language can be automatically detected. Users can save translations in a "phrasebook" for later use. For some languages, text can be entered via an on-screen keyboard, via handwriting recognition, or via speech recognition.

Google Translate is available as a downloadable extension for some web browsers. In February 2010, Google Translate was added to the Google Chrome browser by default for optional automatic translation of websites. In this extension, we can select text on a website and see its translation, and it can also translate an entire page of a website in real time.

4.3.6.3 RapidMiner Studio Software

RapidMiner is a data science software platform developed by the company of the same name. It provides an integrated environment for data preparation,

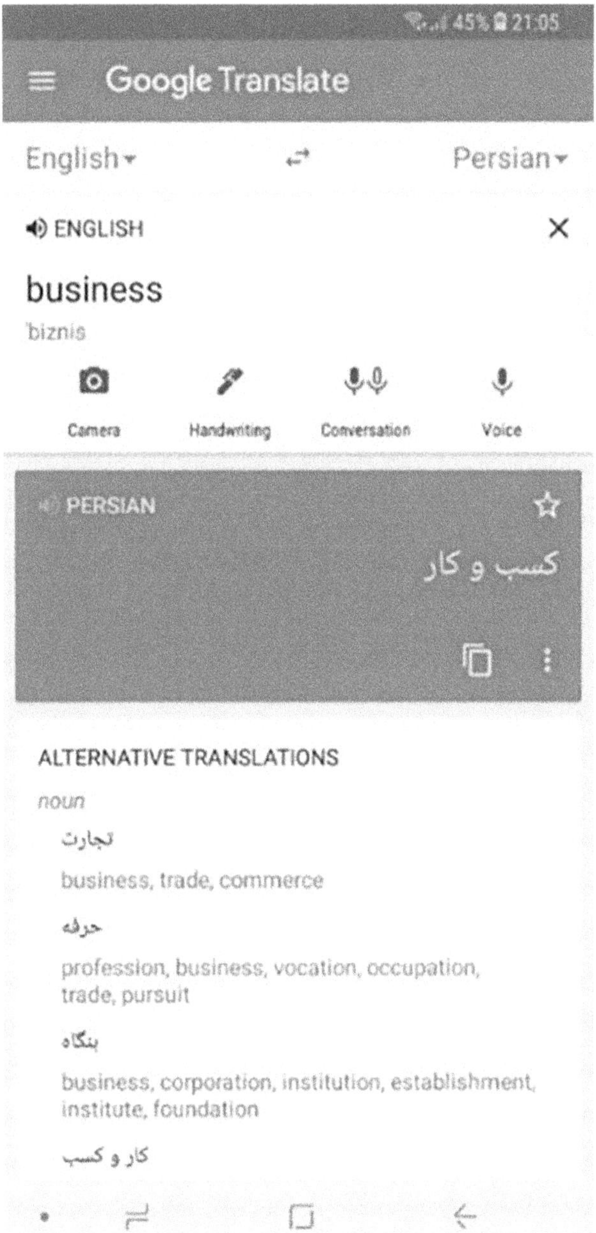

FIGURE 4.3 Google Translate app.

machine learning, deep learning, text mining, and predictive analytics. It is used for businesses and commercial applications, as well as for research, education, rapid prototyping, and software development, and supports all stages of the machine learning process, including data preparation, results visualization, model validation, and optimization. RapidMiner is developed in an open-source model. RapidMiner Studio Free Edition, which is limited to 1 logical processor and 10,000 data rows, is available for free under the AGPL license. Commercial pricing starts at $2,500 and is available to developers.

This product was known as YALE (Yet Another Learning Environment) in its early years, in 2001, and was developed by Ralf Klinkenberg, Ingo Mierswa, and Simon Fischer in the Artificial Intelligence Department of the Technical University of Dortmund. In 2006, it was developed by Rapid-I, a company founded by Ingo Mierswa and Ralf Klinkenberg in the same year. In 2007, the name of the software was changed from YALE to RapidMiner, and in 2013, Rapid-I was rebranded as RapidMiner. This software has attracted the attention of many researchers and students in recent years due to its simplicity of operation, and on the other hand, the open-source development model of this product has also made programmers more willing to develop it. This software uses various algorithms for preparing and modeling information, and this wide range of selectable options has caused many academic projects to use this powerful product, at least for parts of their work. The graphical environment of this program is well-designed. And beautifully designed, users without programming knowledge can easily interact with different parts of the program. The tool provides users with the ability to connect to different programming languages such as Python and R. Even users of the Weka tool are no exception and can connect to this tool. RapidMiner's capabilities can be extended with additional plugins that are available through the RapidMiner Marketplace. The RapidMiner Marketplace provides a platform for developers to create data analysis algorithms and publish them in the community.

RapidMiner is written in the Java programming language, which provides a GUI for designing and executing workflows. These workflows in RapidMiner are called "processes" and consist of several "Operators". Each operator in the process performs a single task, and the output of each operator forms the input of the next operator.

4.3.6.4 MATLAB Software

MATLAB is a fourth-generation high-level programming language and an interactive environment for numerical computation, visualization, and programming, which was created by combining the two words MATrix (matrix) and LABoratory (laboratory). This name indicates the matrix-oriented approach to programming, in which even single numbers are considered as a 1 x 1 matrix. MATLAB software is produced by MathWorks. The company was founded in 1984 in Massachusetts, USA. In 1970, Cleve Moler, the president of New Mexico College, wrote MATLAB software based on the Fortran language. In 1983, this software was based on the C programming language, and its development accelerated after the company was founded.

MATLAB provides the ability to work with matrices, plot various functions and data, implement various algorithms, create user interfaces, communicate with programs written in other languages, including C, C++, JAVA, and Fortran, and create models and applications.

The MATLAB system is made up of five main parts.

MATLAB Language: MATLAB is a high-level matrix-array language that includes object-oriented programming features that can be used to create simple and complex programs.

MATLAB Workspace: A set of tools and features that you, as a MATLAB user or MATLAB programmer, interact with. This environment includes features for managing variables in the workspace and tools for developing, managing, debugging, and creating M-files in MATLAB programs.

Graphics Control: This is the MATLAB graphics system that includes high-level commands for 2D and 3D data visualization, image processing, animation, and graphics. It also includes low-level commands that allow you to customize the graphical appearance of your programs.

MATLAB Mathematical Function Library: Includes a wide range of computational algorithms, ranging from basic functions such as sine, cosine, to complex functions such as matrix inverse, matrix eigenvalues, and fast Fourier transform.

MATLAB Application Programming Interface (API): A library that allows you to write Fortran and C programs that interact with MATLAB. This interface includes features such as calling procedures from MATLAB (dynamic linking), calling MATLAB as a computational engine, and reading and writing to mat files.

4.3.6.4.1 MATLAB Advantages

Ease of Use: MATLAB is an interpreted language that allows programs to be written, modified, and created easily in the MATLAB integrated development environment. Since the programming language is easy to use, it is easy to develop new programs.

Platform Independence: MATLAB is supported by many different computer systems. The MATLAB language is supported by Linux, Windows, and Macintosh operating systems.

Predefined Functions: MATLAB also comes with an extensive library of predefined functions that are used for many applications.

Device-Independent Drawing: Unlike many other computer languages, MATLAB has many commands for drawing and imaging. These images and drawings can be displayed on any graphics output device supported by the computer.

Graphical User Interface: MATLAB includes tools that allow the programmer to interactively create a graphical user interface. With this capability, the programmer can design complex data analysis programs in a way that even inexperienced users can easily interact with the program.

4.3.6.4.2 Disadvantages of MATLAB

MATLAB has two major disadvantages. First, MATLAB is an interpreted language and therefore its execution speed is slower than compiled languages. This problem can be solved by properly applying the program structure in MATLAB and using the MATLAB compiler to compile the final program before distributing it.

The second major drawback is cost: a complete copy of MATLAB is five to ten times more expensive than a C or Fortran compiler. And it is cost-effective for businesses.

4.3.6.4.3 Applications of MATLAB

MATLAB is widely used as a computational tool in science and engineering, such as physics, chemistry, mathematics, and all engineering disciplines. Some of the uses of MATLAB are discussed below:

- Signal and communication processing
- Image and video processing
- Control systems
- Test and measurement
- Financial engineering
- Biological computing

The power of MATLAB's mathematical computations

MATLAB is used in every aspect of computational mathematics. Below are some of the mathematical computations in which MATLAB is most commonly used:

- Working with matrices and arrays
- Drawing 2D and 3D graphs and graphics
- Linear algebra
- Algebraic equations
- Nonlinear functions
- Statistics
- Data analysis
- Differential and integral calculus and differential equations
- Numerical computations
- Integration
- Curve fitting
- MATLAB toolboxes

The manufacturer and various groups, including universities around the world and some engineering firms, add application-specific toolboxes to it every year, which have increased its efficiency and popularity. A list of these toolboxes is given below:

- Simulink: a tool for simulating systems in an abstract manner
- MATLAB Telecommunications Toolbox: functions and tools for telecommunications engineering calculations
- MATLAB Control Toolbox: functions and tools for control engineering calculations
- MATLAB Fuzzy Toolbox: functions and tools for fuzzy calculations
- MATLAB Computational Toolbox: functions and tools for numerical calculations
- MATLAB Estimation Toolbox: functions and tools for system estimation in control engineering
- MATLAB Statistics Toolbox: functions and tools for statistical calculations
- MATLAB Data Collection Toolbox; functions and tools for data collection

- MATLAB Neural Network Toolbox: functions and tools for neural network calculation
- MATLAB Image Processing Toolbox: functions and tools for image processing calculation functions and tools
- MATLAB Audio Processing Toolbox: functions and tools for audio processing calculation
- MATLAB Probability Toolbox
- MATLAB Symbolic Calculation Toolbox
- MATLAB Real-Time Workshop Toolbox: functions and tools for real-time systems calculation
- In this research, we use the Neural Network Toolbox in MATLAB to analyze and examine data

4.4 IMPLEMENTATION AND FINDINGS

This section deals with the steps mentioned in Section 4.3 in a practical and detailed manner, and the research processes are also fully explained. Finally, we will describe and analyze the results obtained.

The aim of this research is to analyze and review customer opinions using data mining tools and neural networks, and to prevent customer churn by providing strategic solutions in the online store. The online store under study in this research is Digikala.

The level of interest and comments of customers on the Digikala social network for the products of this online store was reviewed, and then three products were selected based on the highest level of interest and the highest number of comments.

According to the studies conducted on Digikala social networks, they were in a better position to conduct research and study compared to other pages of this store, because this page was related to the discussion of advertisements and amazing offers of goods, and since our goal in this study is to investigate the effectiveness of advertisements on social networks to attract or lose customers, this is why these pages were selected.

Based on the studies conducted on the products offered on this page, we concluded that the products we studied were digital. In terms of interest and number of comments, these products were more popular than other products. The selected products are briefly introduced in Table 4.1.

4.4.1 Extracting Comments of Selected Products

After selecting the desired products, we proceed to the stage of extracting data and customer comments about the three mentioned products. With the Web Scraping technique explained in the previous section, using the Data Miner plugin, three instructions were written in Figure 4.4 for the selected products.

The columns and, in fact, the features extracted from each comment include:

1. The page address of each user (url)
2. The name of the author of the comment (name)
3. The text of the comment (comment)

TABLE 4.1
Selected Products for the Study

Product photo	Product name	Product type	Number of interests	Number of comments
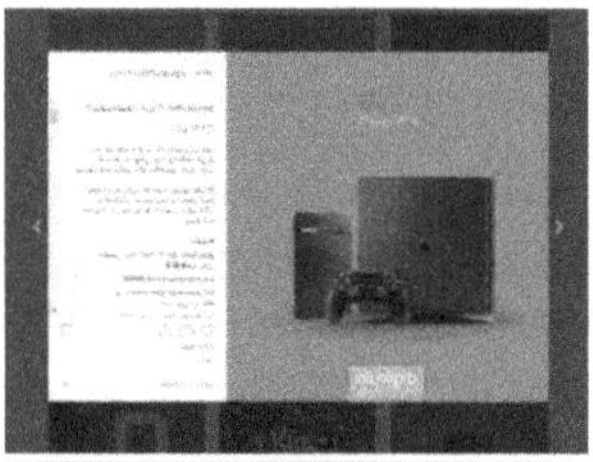	Game console PS5	Digital product	17,800	730
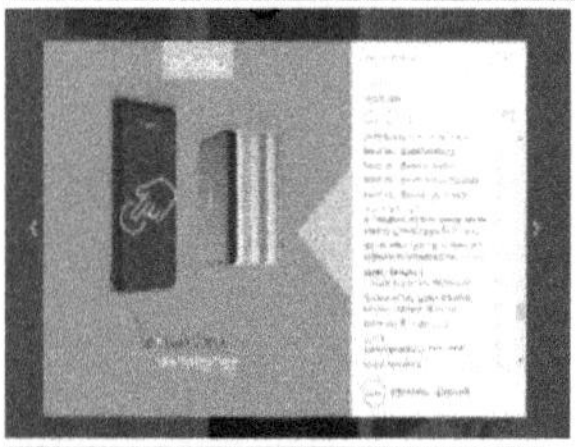	Portable charger	Digital product	18,045	643
	Smart watch	Digital product	13,552	511

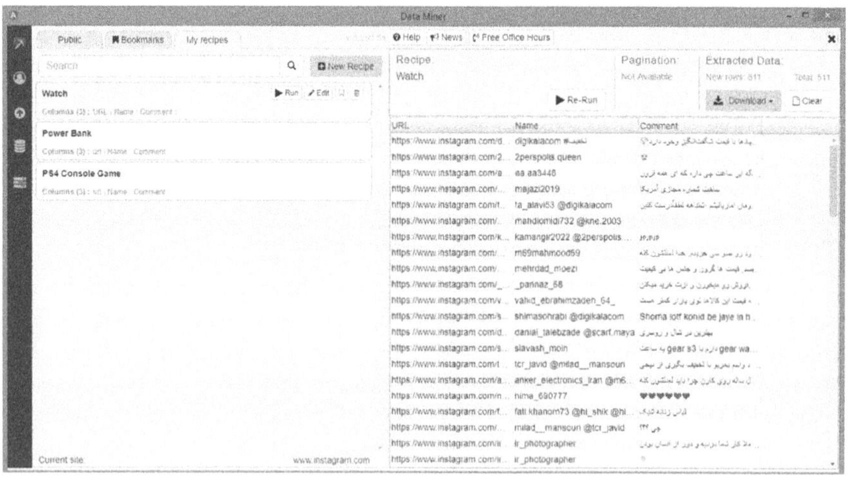

FIGURE 4.4 Data Miner instructions.

4.4.2 Initial Data Preparation and Cleaning

In this section, the URL column, or the address of each user's user page, was changed to ID (of the numeric type).

In the next step, three columns were added to Excel: Price, Quality User Idea, and Like Count, which we will explain below.

The Quality column means the quality of the product from the user's point of view, which is rated based on three criteria: High, which means high customer satisfaction with the quality of the product; Normal, which means average customer satisfaction with the quality of the product; and Low, which means customer dissatisfaction with the quality of the product, which is rated based on users' opinions.

The Price column means the purchase value of the product relative to the price of the product, which is rated based on three criteria: Good, which means high customer satisfaction with the price of the product; Normal, which means average customer satisfaction with the price of the product; and Bad, which means customer dissatisfaction with the price of the product, which is rated based on users' opinions.

The User Idea column means the customer's positive or negative opinion about the store's promotional product, which is based on three criteria: Positive, Negative, and Normal.

The Like Count column is related to the likeability of a comment from the perspective of other users. This feature is related to the Instagram social network, where users can approve that comment by pressing the heart icon next to the user's comments. Here, the Like Count column shows the number of approved comments.

After editing the columns, we come to editing the text data, for which we used the editor software. In this part, the following operations were performed:

- Converting colloquial texts to formal (you cannot←you cannot, because←for, big←big, who←expensive expensive)
- Spelling errors
- Correcting words that interfered with the translation (aaaaa RM Environment and Research Processes)
- Edit markup
- Convert spaces to semi-spaces in two-part words and verbs (I go ← I go, Digikala ← Digikala, Amazing ← Amazing)

4.4.3 RM Environment and Research Processes

After initial data preparation and cleaning, we enter the RapidMiner software environment, or RM for short.

RM includes two important design sections where processes are formed, and the results display the results of executing the processes.

Data integration and initial preprocessing of the merged data set in RapidMiner software

After initial preparation and cleaning of the tables, we transfer the data to RapidMiner software to run the research processes. Our data consists of three products. At this stage, three Excel files, which are our data sets, are added to the RM repository from the add data section. During the data import operation, we change the role of the ID column from Integer to ID as shown in the figure, considering that it has a unique value for each row and that the ID column does not affect the results in data mining calculations. Then we merge the product data with each other through the Append operator.

In this process, first the three desired data sets were connected to the Append operator through the Sub-process operator and the Select Attribute operator, and through this operation, the three desired data sets were merged together and finally saved in an Excel file to be used in subsequent processes.

After merging the data sets in the RM software, it is time to preprocess the data. Some columns have Missing Values or missing data that we must eliminate through processes. In the first step, we temporarily remove the Like Count column, which has a lot of missing data, from the data set using the Select Attribute operator because when the data set has a lot of missing data, the analysis result is incorrect.

In this step, we use the Filter Examples operator to clean this data. In this section, we change the condition class from the parameters section to No_missings_attribute. By performing this process, all missing data is cleaned.

After completing the preprocessing and data preparation steps, we save the final data in an Excel file for subsequent processes.

Converting the data format to Binominal and analyzing the input with the FP-Growth algorithm

To perform the next process, which is analyzing the research inputs using the FP_ Growth algorithm in association rules, we need to convert the data type from nominal to binominal.

Binominal stores data with two values (zero and one or true and false). To perform this operation, we use the Nominal to Binominal operator.

In this operation, except for the Content and Name columns, the rest of the columns are converted to Binominal format.

In the next step, we use association rules to analyze the inputs. To perform this process, we use cleaned data in Binominal format.

At this stage, we need to pass the data through the FP-Growth operator so that we can generate association rules for the selected data of the problem using the Create Association Rules operator.

In this operation, we set the min-confidence value to 0.8. Converting data to numerical format and clustering data with the K-means algorithm: As mentioned in the previous sections, the purpose of clustering in this study is to place data that are more similar to each other in a cluster. Using the clustering algorithm increases the processing speed. In this study, we use the K-means method for clustering. We perform this process using the RM software. In the first step, we need to change the data set from nominal format to numerical format.

The reason for converting the format is that the data must be defined numerically to be read by the next operator, which is clustering. In the next step, as mentioned, the Clustering operator is used for clustering. Since we had determined three features in our data set, we chose the value of k, which is the number of clusters, as three, and the metric type as Euclidean distance.

After clustering the data, we save the output of the clustered data in the RapidMiner software in an Excel file, then we import the data into the MATLAB software for the neural network process.

MATLAB environment and neural network process:

As stated in Section 4.3, we use the neural network toolbox for this research. In this software, the existing toolboxes can be found in the Apps section. For this research, we use the Neural Net Fitting toolbox, which includes the MLP algorithm, and the BP algorithm is used to train the data. Figure 4.5 shows the overview of this software.

In the next step, we must add the input data set of this research, which was preprocessed and clustered in the previous steps, to the software from the Input section. Then, we add the target data set, which is of binary type (due to easier analysis), to the software from the Target section.

Because our data set has many columns and few rows, we select the Matrix Columns option.

In the next step, we need to specify the percentage of training data, test data, and validation data.

The training data is randomly selected by the neural network software, and after training, it validates the data. Next, the test data is compared with the target data for testing.

In this step, the number of neurons in the hidden layer of the neural network is determined by the user. As can be seen in Figure 4.6, the number of neurons is 10.

The data set is trained in this step to obtain the output of the neural network. After training the data and determining the output, the output data is compared with the target data, and then the correlation of the data is measured.

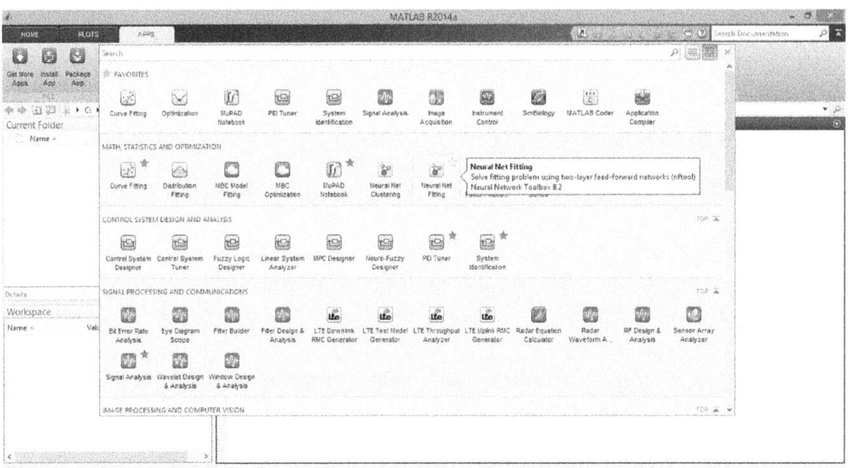

FIGURE 4.5 Overview of the MATLAB software neural network toolbox.

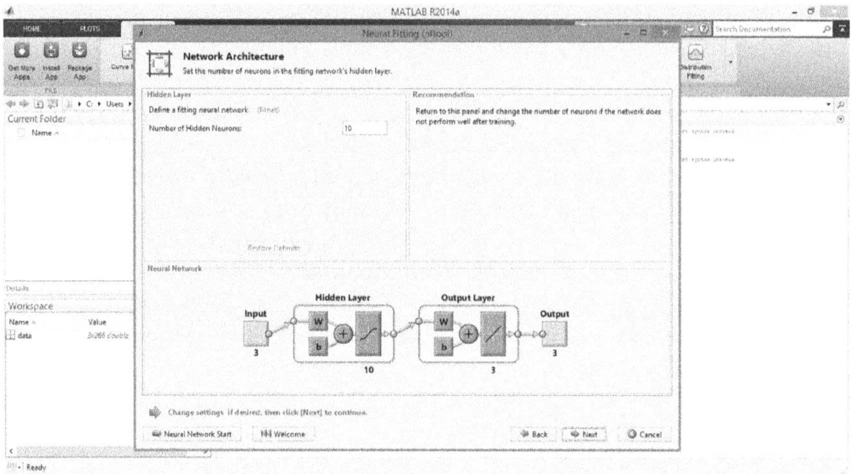

FIGURE 4.6 Specifying the number of hidden layer neurons.

In this study, we have used a neural network that we have written in code to train and test the data set, so that we can perform two other analyses compared to the default analyses of this toolbox.

4.4.4 Description of Results

This section describes the results of the research processes.

Association Rules:

We used the FP-Growth algorithm to analyze the inputs. The result of this process was the production of two rules.

The first rule:

[User idea = Positive] → [Quality = High] (Confidence: 0.846)

The second rule:

[Price = Bad] → [User idea = Negative] (Confidence: 0.955)

The first rule indicates that when users' opinions about the product are positive, the product is likely to be of high quality. The second rule also proves that high prices cause customers to be dissatisfied with the purchase of the product.

Clustering:

According to Figure 4.7, the result of the clustering operation includes three clusters in which the data is scattered based on the ideal distance to the center.

As stated, the Cluster Distance Performance operator was used to analyze the clustering results. This operator shows the average distance within each cluster. The lower the value of these numbers, the more homogeneous the data of that cluster is.

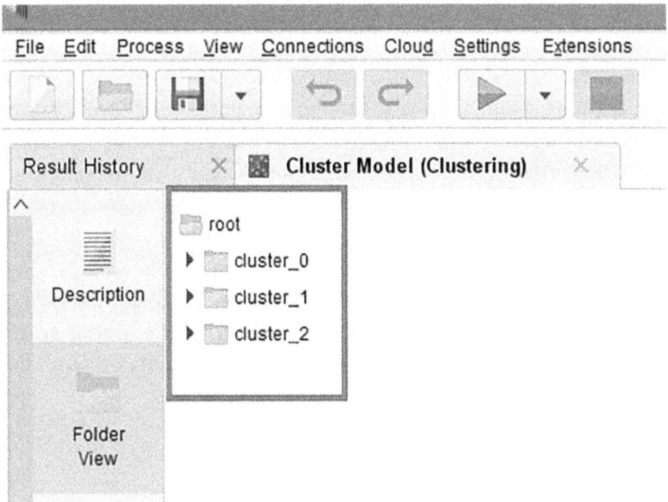

FIGURE 4.7 Result of the clustering operation.

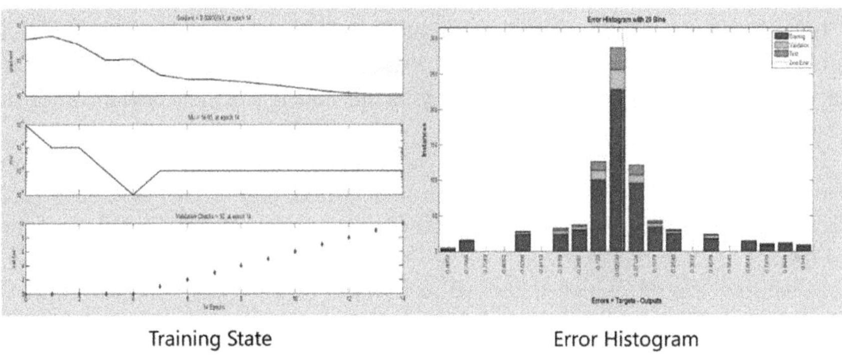

FIGURE 4.8 Error Histogram and Training State graphs.

The Davies–Bouldin index is a measure used to assess the similarity between each cluster and its most similar cluster; the higher its value, the better the clustering is performed.

Neural Network:

After loading the input data and target data and performing the preliminary steps, the input data set, 70% of which were selected for training, was trained using the BP algorithm.

After training the data set, the loop training is stopped after 14 repetitions. Then the results of the Performance, Training State, Error Histogram, Regression, and Confusion matrix graphs can be seen. In Figure 4.8, the Error Histogram and Training State graphs can be seen.

The Error Histogram graph shows the error rate of the neural network output relative to the target data. The trained data is drawn, the validated data is given, and the test data is obtained.

The three Training State graphs show the fluctuations of the Gradient, mu, and Validation Check functions, which have become like this during 14 iterations. These three functions are criteria for stopping the loop. Given that we had considered the Validation Check rate in the neural network code of this study to be 10, the loop stopped after 10 iterations, and there was no improvement.

The regression function graph is shown. Regression indicates the degree of correlation between two target vectors and the output vector of the neural network. The closer the regression value is to one, the more accurate the neural network prediction is. Here, the regression value is 0.78158, which is calculated from the following formula.

$$\text{RE} = \text{Target} - \text{Input}$$

In the Training matrix, out of a total of 212 data that were randomly selected for training, 172 data were correctly placed in their group, and 40 data were incorrectly placed in other groups, so the accuracy rate in this matrix is 81.1%, and the error rate is 18.9%.

In the Validation matrix, out of a total of 27 data that were randomly selected for validation, 23 data were correctly placed in their group, and 4 data were incorrectly placed in other groups, so the accuracy rate in this matrix is 85.2%, and the error rate is 14.8%.

In the Test matrix, out of a total of 27 data that were randomly selected for validation, 23 data were correctly placed in their group, and 4 data were incorrectly placed in other groups, so the accuracy rate in this matrix is 85.2%, and the error rate is 14.8%.

In the final matrix, which is the All matrix, out of a total of 266 data that were selected for training, validation, and testing, 218 data were correctly placed in their group, and 48 data were incorrectly placed in other groups, which is why the accuracy rate in this matrix is 82%, and the error rate is 18%.

The noteworthy point in these four matrices is their second row, which is equal to zero. The reason for this is that the data that have a weaker role in our data set and actually tend to zero are considered by the neural network software to be equal to zero in order to perform a more accurate test.

4.4.5 Analysis of Results

We used the FP-Growth algorithm to analyze the inputs. The result of this process was the generation of two rules.

$$[\text{User idea} = \text{Positive}] \rightarrow [\text{Quality} = \text{High}]\ (\text{Confidence: } 0.846)$$

The first rule indicates that when users have a positive opinion about a product, the product is likely to be of high quality. With this rule, we will conclude that the quality of the product is the first priority for attracting and retaining customers, and a product that does not have the desired quality will no longer be important to most customers.

$$[\text{Price} = \text{Bad}] \rightarrow [\text{User idea} = \text{Negative}] \ (\text{Confidence: } 0.955)$$

The second rule also proves that high prices cause customers to be dissatisfied with the purchase of the product. In addition to the fact that good quality is effective in attracting customers, the price of the product is also of particular importance compared to its quality. According to the second rule, products that are priced high compared to their value are considered unreliable by customers, and this causes customers to turn away from this store and jump to competitors.

In this study, the cleaned data set for stronger and better neural network analysis was first clustered to increase speed and efficiency. For this purpose, the K-means algorithm was selected as a suitable algorithm, which resulted in the division of the data set into three clusters called Cluster 0, Cluster 1, and Cluster 2.

After clustering the data, cleaning was performed again on the data set so that the data could be entered into the MATLAB software environment for analysis. The result of the neural network test in this software was 82% accuracy in predicting whether the user's opinion was positive or negative. This level of accuracy means that new user opinions can be predicted with 82% accuracy so that we can learn about the detection of deviant customers and provide solutions to prevent their loss according to their requests.

In this section, data cleaning was first performed, then the data was entered into the RapidMiner environment for preprocessing. In this stage, the missing data was eliminated. After this stage, the data was analyzed by the FP-Growth algorithm, which is one of the association rule algorithms, in the RapidMiner software, and the result was the production of two rules, which were explained.

Then, the processed data set was clustered into three categories using the K-means algorithm, and the result was saved in an Excel file. In the next stage, the clustered data was entered into the MATLAB software environment, and after applying the neural network test on the data, the output and analysis of the Multi-Layer Perceptron neural network graphs were performed, and the results were explained.

4.5 CONCLUSIONS

As mentioned, online shopping has higher advantages compared to physical shopping in stores, which is why the level of competition among online stores is very high. Given that customers play a key role in these businesses, the need for customer relationship management is felt more than ever. In order to retain previous customers and attract new customers, user opinions and comments require attention, so in this study, we analyzed the comments of users of the Digikala website on the Instagram social network of this website. Initially, different product categories were examined to select the desired product for analysis. The result of the observations was that digital goods were better received for providing opinions. For this reason, we selected three product groups from the digital goods group to analyze and review their opinions. Following the above, three products – PS4 game console, mobile charger, and smart watch – were selected from the digital goods section. Then, the comments of users on the Instagram social network for these three products were extracted using Data Miner software and saved in an Excel file. Analysis of the comments was created in the form of processes in the RM software. The achievements of the results of the processes are as follows:

First, we merged the extracted data together, and then we cleaned and preprocessed the data set related to the three products and saved the output in an Excel file. Before entering the RM software, the data was prepared in several columns through the scoring feature based on three criteria: price, quality and the number of user-liked comments as positive, negative, and normal.

In the second step, to perform the association rules operation, we first converted the data set from nominal to binominal format and then performed this operation using the FP-Growth algorithm, which resulted in the production of two rules.

By analyzing the comments and generating these two rules, we came to the conclusion that in order for the Digikala website to always keep its customers satisfied, it must offer high-quality products, and the prices of the products it specifies for sale on its website must be satisfactory in relation to the value and quality of that product.

In the next step, we needed to cluster the desired data set. The reason for this is to speed up the process steps. Because by classifying similar data sets, each is placed in a category and is processed and analyzed more quickly. At this stage, the data set was converted to a numerical format. Then, through RM software and the K-means clustering algorithm, the data was clustered into three categories: cluster0, cluster1, and cluster2. At the end, the cluster column was added to the data set, and the output was saved in an Excel file.

After clustering the data, it was time to process and analyze the data through neural network algorithms using MATLAB software. We have used the Multi-Layer Perceptron neural network algorithm to perform this operation. To conduct this experiment, we first created training, validation, and test data through sampling. Then, after determining the number of neurons and layers of the neural network, we conducted this test, which resulted in an accuracy of 82% in predicting whether users' opinions were positive or negative. This level of accuracy means that new user opinions can be predicted with 82% accuracy to learn about the detection of deviant customers and provide solutions to prevent their loss in line with their requests. This study can be used as a support for management decisions in website design and product introduction, as well as the use of suppliers in line with customer interests. User opinions on social networks are considered very valuable information for improving online businesses and meeting customer needs. Therefore, conducting various studies to analyze this data into knowledge to predict customer demands from the business in question is essential and important. The following are suggestions for future research in this area:

- Using a database with a larger data volume for better prediction accuracy
- Using a larger number of products with the same categories as well as analyzing non-digital products

- Predicting future customer purchases using association rules by examining customer purchase histories
- Predicting customers who have high lifetime value and trying to retain and maintain them
- Using more optimized and different algorithms with this experiment
- Reviewing user comments on other social networks and drawing conclusions from those comments
- Examining the number of sales after and before the advertisement in a specific time period

5 Text Mining for Loyal Customer Relationship Management

5.1 INTRODUCTION

Although the name e-commerce is becoming a cliché, it is gaining more and more influence in our daily lives. In the past few years, online shopping in Iran was just a dream that was seen only in advanced countries, but due to the increase in Internet penetration in Iran, we are facing the growth and prosperity of online businesses. In "online shopping", consumers are allowed to easily purchase their desired goods or services from a website directly with just a web browser. Today, with the intensification of business and business competition, the issue of giving meaning to data and facilitating business decisions is the center of attention of business owners. "Business intelligence" is actually a set of concepts and methodologies that enable better and more effective decision-making based on facts and data available in the organization. Business has no meaning without customers and cannot continue to exist without them.

To stay competitive, organizations need to know their customers better and understand the differences between them. In fact, the interaction of companies with customers is an important principle in business success. The "Customer Relationship Management" system is one of the most common applications of business intelligence. Customer Relationship Management is a way for a company to move toward customers, the goal of which is to better respond to customer needs. Today, customers are considered a valuable asset for organizations and companies, so attracting new customers and especially retaining past customers, in other words, creating a sense of "loyalty" in them, is of great importance. Organizations use data mining techniques to achieve a better understanding of their customers and operations. "Text mining" is a branch of data mining that focuses on unstructured/semi-structured text files, and with its help, a correct understanding of customer needs can be obtained (Mahdavi et al., 2010).

Until a few years ago, the idea of searching for whatever we wanted on the Internet and having it delivered to our doorstep seemed impossible, but this is what Amazon made possible. The development of the Internet has completely transformed the way we shop. Digikala is an active e-commerce website in Iran, which is considered one of the largest online stores in the country.

 DOI: 10.1201/9781003740148-5

In this study, we present the architectural design of a customer relationship management system for this online business. Customer relationship management is actually a business strategy that predicts and manages the needs of an organization's customers. In this study, we combine customer relationship management systems with business intelligence. One of the effective factors in making business intelligence decisions is "customers". We use business intelligence techniques and text mining techniques to process unstructured knowledge, which in this study is considered as texts, using RapidMiner software. We use this technique to analyze and review customer opinions. These opinions are obtained by extracting user comments from the site.

Our focus in this study is on the purchase of products by customers, and the process of producing products is not considered. Finally, to improve the business and consequently increase revenue and sales, we apply the results to various subsystems of this online business. We also provide suggestions based on the results obtained.

Today, the emergence of online stores that offer various goods in the virtual world is increasing, so that the number of online stores is now more than 900. For online stores to continue operating in this competitive business environment, business processes must be carried out in line with customer expectations, which creates customer satisfaction. In the modern economy, the priority of business activities is customer relationship management, which is considered a business strategy. The effectiveness of this customer relationship management system depends on factors such as its ability to collect data from different customers, which is done by extracting their opinions, evaluating this data, which we use text mining techniques in business intelligence, and finally using the results to improve business processes for customer interaction, which is also our goal in this research to gain a larger customer base and retain previous customers (Fazlollahtabar and Ashena, 2013).

Today, customers are considered a valuable asset for online stores. Our goal in this research is to gain more customers and also retain previous customers, in fact, to create loyalty among customers in the Digikala online store. This goal requires the design of a customer relationship management system. We use business intelligence techniques, including text mining, to study and analyze customer needs. The ultimate goal in this research is to increase revenue and sales and improve business in online stores. To achieve this goal, we apply the results obtained from the research to various subsystems of the store in question and provide suggestions for improving them.

In this research, we seek answers to the following questions:

- Does this customer relationship management system lead to gaining more customers and retaining previous customers?
- How to apply customer review and analysis to online store subsystems?

In this study, the following assumptions are considered:

- The product manufacturing process is not considered.
- Our focus is on the purchase of products by customers.
- The results are applied to the subsystems of ordering, quality control, warehousing, support, and transportation.

- User comments include comment text, strengths, weaknesses, and ratings.
- The format for submitting comments is online and web-based.

5.2 BACKGROUND

Today, with the growth of information systems, organizations have turned to the optimal use of information and converting it into useful knowledge to make their business more intelligent. Municipalities are also considered a suitable platform for business intelligence research due to their large volume of data and the complexity of their communication with citizens.

Researchers tried to understand the hidden behavioral patterns of citizens in the field of tolls and renovation, and the factors affecting their good and bad calculations using data mining tools. The initial steps of data collection, storage, cleaning, aggregation, analysis, and retrieval to achieve business intelligence in municipalities were examined, and using the K-means algorithm, customers were classified into five levels of solvency; these five levels include bad, good, very bad, very bad, and slightly good customers. In this way, the most effective solvency factors for future investments were identified, and the possibility of predicting citizens' behavior in this field was analyzed using decision tree and neural network algorithms (Taheri et al., 2014).

Since the popularity and use of social networks are currently increasing, companies are using them to communicate with customers, manage their brands, and increase loyalty and customer satisfaction. Companies can also use social networks to obtain their customers' opinions, desires, interests, tastes, and preferences. Rezaee et al. used text mining as a method to analyze customer opinions and interests in social networks, whose data was related to the Iranian digital products market.

In this study, two separate clustering techniques and association rules were used to analyze Iranian customers' opinions about mobile phones from six different brands (Samsung, LG, Apple, HTC, Sony, and Huawei). The results showed that from the customers' perspective, no mobile brand is superior, and each of them has positive and negative features. However, Iranian customers have positive feedback on features such as design quality, screen, price, sound, image, and camera, as well as negative feedback on the feature of battery life. Association rules were also used to extract hidden relationships among all mobile phones. The results of extracting association rules for mobile phones provided a strategic guide for digital product service providers so that they could involve their customers' opinions in them and increase customer satisfaction. These rules also became a guide for customers' purchases so that they could buy a phone with the features and advantages they wanted, and on the other hand, customers could have a general understanding of the mobile phones that are offered to the market and the products that will enter the market and could easily recognize the advantages and disadvantages of each (Arabshahi et al., 2020).

With the increase in living standards and the change in people's sustainable living patterns, hairdressing services are widely used by people today. In a study, Ting Wei and colleagues adopted data mining techniques combining self-organizing maps (SOM) and K-means methods to apply the RFM (Lag, Frequency, Financial) model to a hair salon in Taiwan to segment customers and create marketing strategies.

RFM analysis is a marketing method to quantitatively determine the best customers by examining how often a customer has purchased (Lag), how often they purchase (Frequency), and how much money they spend (Financial). RFM analysis is based on the marketing principle that "80 percent of your business comes from 20 percent of your customers". In this study, 11 clusters were formed with the SOM technique and then used to apply the K-means technique to the RFM model. The results of the RFM model analysis showed that the customers of the hair salon can be divided into four types of customers, including loyal customers, potential customers, new customers, and lapsed customers, and unique marketing strategies were created for these four types of customers (Fazlollahtabar, 2022b).

A key aspect to maintain a store in the market is the analysis of customer types, in other words, customer typology, which by understanding different segments of consumers, entrepreneurs can achieve great benefits. Researchers proposed a methodological approach of a decision support system to identify types of Internet customers. The added value of the work was that the online customer segmentation process includes predictive mechanisms that significantly expand the practical capabilities of the developed system. Methodologically, the details of a unique approach were demonstrated, which consists of classical customer segmentation methods and graphical automation in one system, thus providing a wider opportunity for collecting practical information (Fazlollahtabar, 2012b).

A conceptual framework was developed, based on which a prototype of the online store was created for testing. Then, a group of randomly selected users was isolated and, based on the collected data, their clustering was performed by cluster analysis, and a typology of online store users was created. In addition, by introducing a prediction mechanism based on the mathematical model of cellular automata graph, a new customer can be quickly added to the defined group of store customers. The results of this research included a logical division of customers into four groups: opportunity hunters, new product followers, customers eager to visit online stores, and information hunters (Fazlollahtabar, 2008).

Internet technology enables companies to attract new customers, track their online performance and behavior, and customize communications, products, services, and prices. Analyzing customers and customer interactions for electronic customer relationship management (e-CRM) can be done using data mining, optimization methods, or a combination of methods. A key issue in analyzing web access patterns is the clustering and classification of web documents. Generally, classification is based on analytical models that assume a set of predefined keywords with a predefined list of categories, which is not realistic for large and evolving document collections such as the World Wide Web (Fazlollahtabar and Aghasi, 2012).

Researchers proposed a neuro-fuzzy approach to solve the problem of an infinite number of evolving categories that automatically place web documents into different categories. The aim of this approach was to facilitate the preprocessing of clickstream data in e-CRM, which included dynamic changes in web documents. The approach started by classifying test documents into a set of initial categories. A customer satisfaction page was designed to collect customer ideas about products. After collecting customer information, clusters were created based on the DWDC algorithm. Then, a prototype system based on fuzzy clustering of customer relationship management

(FC-CRM) was developed to validate the proposed approach and demonstrate how it handles the dynamic incoming flow of new documents. The designed FC-CRM software was capable of performing clustering, building user profiles, statistical analysis, creating graph features, etc. Based on the logic proposed in this study, adaptive interaction with customers can be automated, which will help provide a "self-active" process of digital content services (Mehrmanesh et al., 2013).

The customer is considered a vital asset of the company. It is very important that the customer relationship management decision-maker summarizes the available information to make the decision-making process fast and efficient. Researchers examined a questionnaire consisting of nine categories of satisfaction variables to obtain the required information. Using principal component analysis (PCA) and data envelopment analysis (DEA), the analysis was performed, and PDA was used as a combination of the integration of these two methods. In this study, an intelligent customer relationship management framework was proposed to improve overall customer satisfaction. The approach to building the CRM framework consisted of five steps (Fazlollahtabar et al., 2010).

To analyze the method effectively, PCA was used to assign a number to each set of questions related to each satisfaction variable, and to achieve the desired accuracy, DEA was applied to three customer categories (most important customers, important customers, and regular customers) to determine the strengths and weaknesses of customer service from the customers' perspective. Customers were then clustered and then DEA was used to determine their perspectives. Using DEA, they optimized their customer feedback recognition and then provided recommendations and corrective actions to solve current issues in the field of logistics and public transportation, especially in the Port of Fremantle.

Also, software calculations and optimization techniques were integrated to build the proposed CRM system. The importance of the proposed algorithm in this study is threefold because, first, it integrates software computation and optimization techniques to build the proposed CRM system; second, it uses most of the standard CRM variables in the decision-making process; and third, it is an optimization algorithm because it integrates DEA with PCA techniques.

Today, complaints and complaint handling play an important role in customer relations. Although companies are reluctant to say that they have a problem with customer complaints, it still seems that they have not found a complete solution to this issue. To rectify this situation, customer complaints should be comprehensively collected and analyzed, problems should be categorized, and solutions should be developed in a timely manner.

In this regard, researchers conducted an analysis of the relationships between the main components of customer relationship management (CRM) and customer complaints in the field of logistics and transportation. In this study, a conceptual framework was presented including mathematical models, hypothetical relationships, perceived value, and interaction between customer, business, and system, as well as customer satisfaction analysis. This framework estimated the relationship between customer satisfaction, loyalty, and acquisition problems and evaluated customer satisfaction and loyalty. Both qualitative and quantitative methods were used for analysis. A survey form was also distributed among 60 customers of Fremantle Port in transportation and logistics to collect data. For the quantitative approach, linear and nonlinear modeling was adopted. Using this model, they were able to address the weaknesses of CRM technology and solve the problems related to improving loyalty and acquisition. To achieve the research objectives, the problems were resolved using a four-step method: text mining analysis, data coverage analysis, linear modeling, and adaptation to a network-based fuzzy inference system. After reviewing multiple tools and methodologies to test and validate all hypotheses, the adaptive fuzzy inference system (ANFIS) was selected as the appropriate tool to prove the relationships. There are different types of FIS. TS-FIS is a systematic approach that obtains an output from a set of inputs, where the relationship between the output and the inputs is defined by fuzzy if-then rules, which was proposed in this study (Taherikhonakdar and Fazlollahtabar, 2024).

Finally, based on nonlinear modeling and using a fuzzy inference system, called the Takagi–Sugeno-type method, fuzzy rules were defined, that is, the relationship between customer satisfaction and the main relevant variables was determined. The results of linear analysis showed that although perceived value and interaction can create positive satisfaction, they do not have the effective ability to generate sufficient loyalty and customer acquisition. For this purpose, by using appropriate rules, the fuzzy logic method was used to evaluate and predict loyalty, customer acquisition, and customer satisfaction, and the conceptual framework in this study could be applicable to any business environment. The fuzzy logic approach also created rules for handling complaints and increasing each of the applied variables.

As online shopping is growing, the probability of service failure is also increasing. Therefore, online retailers need to use specific recovery strategies in their online settings. Online shopping relies more on information and communication than traditional shopping. Furthermore, involving consumers in the recovery process always makes their satisfaction after recovery higher than usual and hence, behavioral decisions such as positive word-of-mouth, repurchase intention, loyalty, and intention to establish cooperation in the future increase (Fazlollahtabar, 2024).

While recent research in online shopping has not examined the effects of informational justice and customer cooperation on retrieval, Gohary et al. conducted a scenario-based experiment between subjects to investigate the effect of four justice dimensions (distributive, procedural, interactional, and informational) on the mediating role of post-retrieval satisfaction to reduce strong memories and subject biases. A sample of 278 Iranian online shoppers on two shopping websites, Digikala and Bamilo, was used for data analysis. Covariance-based structural equation modeling was used to analyze the data.

First, the measurement model was implemented to examine the validity and reliability of the constructs involved in the conceptual framework. Then, the proposed hypotheses were tested using the structural model. The results showed that justice dimensions are positively related to post-retrieval satisfaction. It was also confirmed that the partnership between the dimensions of justice and satisfaction, which means the participation of customers in the recovery process, is essential in the online post-recovery evaluation. The results suggested that managers should generously share their information with customers and explain the reason for the failure in detail to them.

According to Seoul Financial Internet News in 2008, South Korea ranked first in terms of the percentage of Internet access at home, with 80% of the total population, two-thirds of whom were online shoppers. Many tourism companies are now actively using websites as a key tool for marketing and selling their products and services, so to be successful, tourism e-commerce services must be reliable.

With the rapid growth of the Internet, the ability of users to create and publish content has created active electronic communities that provide a wealth of information. However, the sheer volume of reviews and opinions that are typically published for a product makes it harder for individuals and manufacturers to find the best reviews and understand the true quality of the product. Researchers conducted research that re-examined the impact of reviews on economic outcomes, such as product sales, and also examined how different factors affect social outcomes such as perceived usefulness. The research approach examined various aspects of the review text, such as subjective levels, various measures of readability, and the amount of spelling errors, to identify important text features.

Econometric analyses showed that the extent of subjectiveness, informativeness, readability, and linguistic accuracy in reviews significantly affected sales and usefulness. Reviews that are a mix of objective and highly subjective statements have a more negative impact on product sales than reviews that are purely subjective or objective. However, such reviews are more likely to be perceived as helpful by other users. In fact, this study, using random forest-based classifications, showed that it can accurately predict the impact of reviews on sales and their perceived usefulness. It also examined the relative importance of three broad categories: reviewer-related features, subjective review features, and review readability features, and found that each of these three features can yield statistically equivalent performance results. This study is the first to combine econometric methods, text mining, and predictive modeling techniques to conduct a more comprehensive analysis of the information captured by user-generated online reviews to assess their usefulness and economic impact.

Technological advances have made it possible to ubiquitously disseminate user-generated content, including reviews of products that are publicly available on the Internet. User comments and reviews on e-commerce websites such as Amazon.com have a significant impact on the reputation of products, as prospective buyers heavily research them before making a purchase decision.

Past researchers presented a framework using text mining to analyze consumer behavior and collect customer feedback on Amazon using R software. Any individual or business can use this framework to gain insight into what customers say about the products they purchase and customize the methodology to suit their needs. The case study in this research involves comparing and contrasting the positive and negative aspects mentioned by users of three devices (two types of tablets and one laptop) which is useful for improving future generations of products.

By understanding all the positive and negative aspects of a product, manufacturers can align with the acceptance of their products. It also enables them to identify problems revealed by customer reviews and comments and address and resolve these problems. As a result, manufacturers were able to review and incorporate product features when a product was launched and, over time, correct product problems, understand customer requirements, and establish customer satisfaction.

Dixit and Kr presented a collaborative analysis of customer feedback from a specific brand, and to obtain this review, they collected customer reviews and feedback from an e-commerce website.

In this study, customer feedback analysis was performed on three different mobile phone brands: Nokia Lumia, Sony Xperia, and Samsung. Two different classifications were created to extract customer feedback, which were shared on an e-commerce website and divided into three categories (good, bad, and mixed). In this study, two classification algorithms were used: Naive Bayes and k-NN. The accuracy and recall of both classifications were compared, and then the classification that had the best result in the given conditions was obtained. These reviews were in the form of English vernacular; Therefore, to obtain some valuable information from these studies, a text mining system was used, in which the RapidMiner tool was used in this study to classify and prepare the dataset for testing.

Considering recent work in this study, we present the architectural design of a customer relationship management system for an online business, which is a digital goods online store. Then, by using business intelligence techniques, including text mining, with the aim of acquiring a larger customer base and also retaining previous customers, in fact, creating loyalty in customers, we analyze and review customer opinions obtained through site comments using RapidMiner software. And finally, to improve the business and consequently increase revenue and sales, we will apply the results to the various subsystems of this online business and provide suggestions.

5.3 PROBLEM AND METHODOLOGICAL FRAMEWORK

According to the contents of the previous sections, first we explained the generalities of the research and then we talked about the literature and background of the research, which led us to a problem. Our research problem includes details that we are going to explain here. Finally, we will describe the tools used in the research.

With the advancement of technology, people are looking for the most convenient ways to do their daily tasks, including online shopping. There are many reasons to shop online instead of going to stores. Online shopping is more convenient, meaning that it is always available and possible 24 hours a day, 7 days a week, and in any weather conditions without having to look for a parking space and endure the traffic jams of these days. Other reasons include the ease of searching for desired products, having a variety of choices from different brands, and the ability to compare products and their prices with each other and reach a good decision to buy. Other benefits of online shopping include saving time, energy, and fuel. As a result, online shopping has become the first choice of most people, and most consumers prefer online shopping, where they can buy what they need with a few simple clicks, and all they need to do is wait for it to be delivered to the desired address, which may even be a long distance from them.

So, considering the advantages of online shopping, we can conclude that the competition between online stores is very high, and customers play a key role in this business. In fact, the online selection of products is up to consumers and customers; they have the power of choice and the world of online retailers at their fingertips. Meanwhile, if the online store does not meet their needs, they quickly go to a competitor's store. This is where the big issue of customer loyalty arises, which is also the issue we are discussing in this research: increasing loyalty in online businesses. Increasing customer loyalty plays an important role in the success of the organization, especially when customer purchases are not equivalent to long-term success. The concepts of satisfaction and loyalty to a business website are the main concerns of marketers. E-loyalty is the customer's favorable attitude toward the online store website, which forces the customer to repeat the purchase behavior. A loyal customer always intends to buy from a specific website and does not move to another website, which is a great competitive advantage. Keeping an existing customer is cheaper and more economical than searching for a new one. Loyal customers are more likely to spread favorable messages about positive past service experiences and recommend the company to new and non-loyal customers, creating the possibility of increased advertising at no additional cost to the service provider. Therefore, a loyal customer is considered a major asset for online businesses. We conclude that e-loyalty is a strategic necessity for long-term business success and, in general, customer retention is directly related to the profitability of online retailers, so investing in this area will be effective and beneficial.

This is where having a customer relationship management system becomes necessary to align business processes with customer expectations and needs and to run the organization around the customer. Retailers must be responsive to customer needs, manage and anticipate them. In this study, we present the architectural design of a CRM system for intelligent customer use.

In this study, we analyze the opinions, reviews, and ratings of customers and consumers who have shared their experiences of purchasing products online on a store website to measure customer loyalty in a survey mechanism. Product reviews and ratings on a website are an important source of information for customer decision-making. According to the research conducted, it has been found that about 90% of users read reviews, and more than 80% of them trust the opinions of consumers who are considered strangers, which is surprising. In fact, users are interested in these reviews and opinions because they provide third-party insight into the quality and experiences of their product. So, these reviews and opinions are very powerful, and when used effectively, they can help the store increase sales. It is clear that ignoring these opinions by the organization as a marketing opportunity is considered to be ignoring a large percentage of their buyers. Reviews and opinions can eliminate any doubts that customers have about a product or the choice of that product. Ultimately, it can be said that customers play a very important role in online business, and paying attention to their opinions and reviews is also considered very important.

Making sense of, collecting, maintaining, and organizing data in organizations play an important role, where most of our data here consists of customer opinions and reviews, which are unstructured texts. So, in this study, we use text mining techniques. We will discuss the text mining process used in this study and its algorithm in the next steps. Next, using these techniques, we will examine consumer opinions and create processes with the goals of finding active users, examining user types and their suggestions, analyzing the strengths and weaknesses of products, and classifying with k-NN and Naive Bayes algorithms in RapidMiner software.

5.3.1 Research Framework

In the process of purchasing from online stores, the user searches and examines different sections of a website to find the product they want. In this process, the user will be faced with various and diverse options and may easily pass by a number of high-quality products due to the lack of comments. In this research, finding the desired product for analysis is considered to be one of the initial and important stages. An attempt has been made to use the best-selling, most visited, and most popular products that also have many comments at this stage.

After reviewing and selecting the desired data, we reach the next stage, namely extracting and translating them. Web mining is the application of data mining techniques to identify patterns on the web. Web mining is divided into three categories: web content mining, web application mining, and web structure mining. Web content mining is the process of extracting useful information from the content of web documents. The content of a web document corresponds to the concepts that the document is trying to convey to users. This content can include text, images, video, audio, or structured records such as lists and tables. Among them, text mining has been researched more than any other field. In this research, we use the web content mining method to extract comments and take help of the Web Scraping technique (explained in detail below) using the Data Miner extension in Google Chrome. Google Translate is also used to translate the comments simultaneously.

5.3.1.1 Web Scraping

Web Scraping is a technique for extracting large amounts of data from websites, where the extracted data is stored in a local file on your computer or in a spreadsheet format in a database. This technique is also known as Screen Scraping, Web Data Extraction, and Web Harvesting.

The data displayed by most websites can only be viewed using a web browser. They do not provide a function to save a copy of this data for personal use. The only option is to manually copy and paste the data, which is a very tedious task and can take several hours or sometimes several days to complete. Web Scraping is actually the technique of automating this process, so that instead of manually copying data from websites, Web Scraping tools do the same thing in a matter of seconds. Web Scraping tools automatically load and extract data from multiple website pages based on your needs. It is either custom-built for a specific website or can be configured to work with any website. With the click of a button, you can easily transfer the information from the website to a file on your computer. There are many software tools that can be used to customize Web Scraping solutions.

After Web Scraping, the extracted data is saved as tables in Excel. Then we need to clean and prepare this saved data using Excel and the editor plugin. With a series of initial preprocessing steps that will be explained in detail in the next section, we prepare the data for entry into the RapidMiner software. After these steps, the data enters the RapidMiner environment. Processes and procedures for analyzing and reviewing comments are formed. In processes that use text data, because this data is unstructured, it requires preprocessing. Preprocessing at this stage includes a series of text mining operations. First, the texts are separated (Tokenize). Then, stop words in Persian and English are removed from the text. All uppercase letters are transformed to lowercase letters (transform cases). After that, stemming is performed using the Porter method. Then, trigrams and 4-grams in the text are created, and finally, words of the desired type, for example, adjectives, are extracted by POS tagging. In data mining processes that are performed for prediction, we need to convert text data into vectors and numerical data. This is done with the TF-IDF method, which is described below.

5.3.1.2 TF-IDF Method

The TF-IDF method, which stands for term frequency–inverse document frequency, is a statistical measurement and counting index in the field of language and speech processing that is often used as a weighting factor in text mining and information retrieval. The purpose of this index is to determine the importance of a word in the text, which it does by giving less points to less frequent words and more points to more frequent words, which also helps to control the fact that some words are more common than others. For example, when entering an address, there is a possibility that the word ally will be recorded instead of the word ally. Therefore, the matching criterion ignores less important words such as ally, alleyway, and street and pays attention to more important words in the text. In this method, we consider the frequency of a word in a document against the number of times it occurs in the set of all documents. The weighting of words is a function of the distribution of different words in the documents. To implement this method, we first consider a set of documents. For all the words in the corpus, we check how many documents each word has been repeated in and save it. Then, a document is received as input. First, we check how many times each word in the input document has been used in the same document. Then, for all the words in the input document, we check how many documents each word appears in. The weighting of words is determined using two criteria: term frequency and inverse document frequency, which are calculated as follows:

Calculating TF according to Equation (5.1) is simply the ratio of the number of times a keyword appears in a given document, n_k (where k is the keyword), to the total number of words in the document, n:

$$TF = n_k / n \tag{5.1}$$

For example, a common English word like "that" will have a relatively high TF value, and a word like "RapidMiner" will have a much lower TF value. The IDF is also defined according to Equation (5.2):

$$IDF = \log_2 \left(N / N_k \right) \tag{5.2}$$

where N is the total number of documents being examined. For most text mining challenges, N is the number of documents we are trying to mine, and N_k is the number of documents that contain keyword k.

A word like "that" will certainly appear in every document, so the ratio (N/N_k) will be close to 1, and the IDF value for "that" will be close to zero. Whereas, a word like "RapidMiner" will probably appear in a relatively smaller number of documents, so its ratio (*N/N*k) will be much greater than 1. Therefore, the IDF value will be higher for keywords that are less common. Finally, the weight of a word, or TF-IDF, is calculated as in Equation (5.3).

$$TF-IDF = n_k/n \times \log_2\left(N/N_k\right) \tag{5.3}$$

According to the example, when we multiply the high TF for "that" by a very low IDF, we get a very low (or zero) TF-IDF, while multiplying the low TF for RapidMiner by a relatively high IDF will get a relatively high TF-IDF. Typically, TF-IDF values are calculated for each word in a document set during the preprocessing stage of the text mining process. In the RapidMiner document processing operator, we have four optional parameters to create the vector, including the following:

Binary Term Occurrence: It places the value 1 in the intersection cell between the document (rows) and the token (columns) if the token occurs at least once in that document, and the value 0 otherwise. The number of occurrences in the document is ignored in this scale.

Term Occurrence: It places the exact number of occurrences of each token in the intersection cell. If the word or term of interest does not exist in that document, the value 0 will be placed in the intersection cell.

Term Frequency: It places the relative frequency of the token in the document in the intersection cell. This measurement is calculated by dividing the number of occurrences of a word/term (Term Occurrence) by the total number of words in that document.

TF-IDF: Arguably one of the most common numerical representations of text used in text mining. TF-IDF calculates a numerical value that emphasizes both the frequency of a token in a document and the rarity of that token in the set of all documents discussed above.

In this research, we use two values, TF-IDF and Term Frequency, in text mining calculations.

5.3.2 Algorithms

Now that our data has been preprocessed, it is ready to be used in RapidMiner processes and data mining algorithms. In this study, about eight processes are written, which will be analyzed in detail in the next section. Some of the processes include classification operations and predicting features in products. In these processes, two algorithms are used, k-NN and Naive Bayes. The description of these algorithms is given below.

5.3.2.1 Bayesian-Aided Classification

One of the important probability formulas is the Bayesian probability formula, with the help of which we can estimate the probability of a class label of a data sample.

Using this rule for classification provides good accuracy and speed in large databases. Bayesian theorem is named after an English clergyman named Thomas Bayes, who did work in the field of probability and decision theory in the 18th century. Suppose X is a tuple and is usually described by the values of a set of n attributes. Suppose H contains some assumptions, such as the tuple X belongs to a certain class, such as C. Here, we want to determine P(H|X), the probability that the assumption H is held by the observed tuple X. In other words, we want to find the probability that the tuple X belongs to class C, given the values of the attributes of X. Here, P(H|X) is the posterior probability of H given X. On the other hand, P(H) is the prior probability of H. Similarly, P(X|H) is the posterior probability of X given H. The expression P(X) is the prior probability of X. The probabilities P(H), P(X|H), and P(X) can be estimated from the data set. Using Bayes' theorem, the posterior probability P(H|X) can be obtained from the above three probabilities. Bayes' theorem is expressed as in Equation (5.4):

$$P(H|X) = \frac{P(X|H)P(H)}{P(X)} \tag{5.4}$$

Below is an explanation of how Bayes' theorem is used for classification.

We denote the set of training tuples with their class labels by D. We typically represent each tuple by an n-dimensional vector $\boldsymbol{X} = (x_1, x_2, \ldots, x_n)$, where x_i is the value of the attribute A_i.

Suppose there are m class labels named $C_1, C_2, \ldots, C_m$. Consider a tuple like X. The classifier looks for the one with the highest posterior probability given X. In effect, the classifier predicts the tuple X to belong to this class. So, X belongs to class C if and only if we have Equation (5.5):

$$P(C_i|\boldsymbol{X}) > P(C_j|\boldsymbol{X}) \; for \; 1 \le j \le m, \; j \ne i \tag{5.5}$$

So, we are looking to maximize $P(C_i|\boldsymbol{X})$ and class C_i also called the maximum posterior. With the help of Bayes' theorem (Equation (5.4)), we have Equation (5.6):

$$P(C_i|\boldsymbol{X}) = \frac{P(\boldsymbol{X}|C_i)P(C_i)}{P(\boldsymbol{X})} \tag{5.6}$$

Since the value of P(X) is constant for all classes, we only need to know with which of the classes the expression $P(\boldsymbol{X}|C_i)P(C_i)$ has the largest value. If the prior probabilities of the classes are not known, it is common to proceed with the assumption of equal probabilities, i.e., $P(C_1) = P(C_2) = \ldots = P(C_m)$, in which case we only need to seek to maximize $P(C_i|\boldsymbol{X})$. These probabilities may also be estimated using the formula $P(C_i)=|C_{i,D}|/|D|$, where $|C_{i,D}|$ is the number of training tuples labeled with class C_i.

Data sets with a large number of attributes cause heavy calculations for $P(X|C_i)$. To reduce these calculations, the assumption of conditional class independence is used. If we want to express this assumption in simple terms, it is stated that there are no dependency relationships between attributes, and, in fact, the values of the attributes are conditionally independent of each other. So we have Equation (5.7):

$$P(X|C_i)=\prod_{k=1}^{n}P(x_k|C_i)=P(X_1|C_i)\times P(X_2|C_i)\times\ldots\times P(X_n|C_i) \tag{5.7}$$

With the help of training tuples, we can easily estimate the probabilities $P(X_1|C_i),P(X_2|C_i),\ldots,P(X_n|C_i)$. Note that x_k refers to the value of the attribute A_k in the tuple X. For each attribute, it must also be checked whether it is discrete or continuous. For example, to calculate $P(X|C_i)$ for these two cases, we proceed as follows:

a) If the attribute A_k is discrete or categorical, to calculate $P(x_k|C_i)$ it is enough to divide the number of tuples by the class label C_i whose value of the attribute A_k is equal to x_k by the number of tuples with the class label C_i (i.e., $|C_{i,D}|$).
b) If the attribute A_k is continuous, our work will be a little more, but the way is very simple. It is usually assumed that a continuous feature has a Gaussian distribution with mean μ and standard deviation σ, which is represented by Equation (5.8):

$$g(x,\mu,\sigma)=\frac{1}{\sqrt{2\pi\sigma}}e^{\frac{(x-\mu)^2}{2\sigma^2}} \tag{5.8}$$

So, we have Equation (5.9):

$$P(x_k|C_i)=g(x_k,\mu_{c_i},\sigma_{c_i}) \tag{5.9}$$

5.3.2.2 K-Nearest Neighbor (k-NN)

The k-NN algorithm became popular as the computing power of computers increased, and one of its common applications is pattern recognition. For a test data set, the algorithm looks for the k closest examples (similar examples). The proximity of two samples is calculated by obtaining the similarity or distance between the two samples. Each sample can consist of data types that need to be examined for similarity.

After finding these k data similar to the test sample, the test data class label is selected by majority vote. If the value of k is set to 1, then the class of the data closest to the test sample is presented as the estimated class. However, due to the presence of noisy data and out-of-range values, the value of 1 is not a suitable number for k. The appropriate value can be obtained empirically. For example, starting with 1 and calculating the error rate for a test set. We repeat this by increasing the value of k. A value of k that causes the minimum error rate is a suitable choice.

This algorithm can also be used for data whose class label is continuous (numeric). In this case, after finding k neighbors, the average of the values obtained from the class of these k samples is selected as the test sample class label. Whenever the value or values of the attributes in the original data set are incomplete with an experiment, the minimum distance for this attribute is considered in the calculation of similarity, and the maximum distance for this attribute is considered in the calculation of distance. Suppose the values of a numerical attribute are mapped to the distance from zero to one. You know that this can be done with a simple normalization. To calculate the distance between two samples and for this attribute, if both values are incomplete, the distance is considered equal to one (similarity equal to zero). However, if one of these values is incomplete and the other has a value equal to v, the distance and similarity are obtained from formulas (5.10) and (5.11):

$$Dis\left(A_i, A_j\right) = Max\left(|1-v|, |0-v|\right) \tag{5.10}$$

$$Sim\left(A_i, A_j\right) = Min\left(|1-v|, |0-v|\right) \tag{5.11}$$

For non-numeric features, it is sufficient for at least one of them to be incomplete to set the distance equal to one and the similarity equal to zero. By assigning a weight to each feature, we reduce or increase the percentage of the feature's contribution to the calculation of the similarity or distance between samples. In this way, irrelevant features or noise data will have less impact on the process.

5.3.2.3 Evaluating the Accuracy of Classifications

This section discusses metrics for evaluating classifiers and examines how they perform in predicting class labels. Using training data to create classifiers and estimate the accuracy of the resulting model can lead to optimistic estimates for evaluating accuracy, because the same training data was used to build the classifier or model; therefore, it is better to use a test dataset to evaluate the model, which has not been used to train the model before.

Before discussing the different metrics, it is necessary to familiarize ourselves with some terminology. In the case of positive tuples (tuples of a class that the user is interested in) and negative tuples (the rest of the tuples), for example, consider a dataset with two class labels. Positive tuples can be considered tuples labeled yes for the class attribute (buys_computer=yes) and negative tuples are those whose class label is no (buys_computer=no). Suppose we run the resulting classifier on a test set of labeled tuples. We denote the number of positive tuples by P and the number of negative tuples by N.

There are four other terms that are actually basic concepts for calculating many evaluator metrics.

- True Positives (TP): This term refers to the positive tuples that are correctly labeled by the classifier. The number of these tuples is denoted by TP.
- True Negatives (TN): This term refers to the negative tuples that are correctly labeled as negative by the classifier. The number of these tuples is denoted by TN.
- False Positives (FP): They are negative tuples that are incorrectly labeled as positive (e.g., tuples with the class label No that the classifier predicted to be labeled as Yes). We denote the number of these tuples by FP.
- False Negatives (FN): They are positive tuples that are incorrectly labeled as negative (e.g., tuples with the class label Yes that the classifier predicted to be labeled as No). We denote the number of these tuples by FN.

The accuracy of a classifier on the test data set is the percentage of tuples in the set that are correctly labeled by the classifier. So, in Equation (5.12), we have:

$$accuracy = \frac{TP + TN}{P + N} \tag{5.12}$$

In the field of pattern recognition, this metric is known as the classifier's recognition rate and reflects how well it distinguishes different classes. We can also calculate the error rate or misclassification rate of the model M from the simple formula 1–accuracy (M). Where accuracy (M) represents the accuracy of the model. This value is calculated from Equation (5.13).

$$error\ rate = \frac{FP + FN}{P + N} \tag{5.13}$$

Precision and recall metrics are also widely used in classification. The precision metric is considered a measure of accuracy (the percentage of tuples that are labeled positively and are actually of positive class), and the recall metric is a measure of completeness (the percentage of positive tuples that are correctly classified). These two metrics are calculated through Equations (5.14) and (5.15):

$$precision = \frac{TP}{TP + FP} \tag{5.14}$$

$$recall = \frac{TP}{TP + FN} = \frac{TP}{P} \tag{5.15}$$

Another solution to use precision and recall is to combine them and convert them into another simple metric. This is the same approach used in the F-score metric. This metric is obtained as in Equation (5.16). In fact, the F-score metric is the harmonic mean of the two metrics: precision and recall.

$$F = F - score = \frac{2 \times precision \times recall}{precision + recall} \quad (5.16)$$

Finally, we will analyze and review the results of the processes to increase customer loyalty, increase the range of customers, and provide various solutions to improve online business.

5.3.3 Tools Used in the Research

This section introduces the tools and software used in this research.

5.3.3.1 Data Miner Extension for Google Chrome Browser

Data Miner is a Google Chrome extension that helps us save information from web pages in a CSV file or Excel spreadsheet. This tool can be downloaded from the site https://data-miner.io or from the Google Chrome browser extensions section.

Data Miner has more than 40,000 general instructions for many popular sites. Using these general instructions, you can get the structured data you need with just one click and save time. Some tools only work on specific websites, but Data Miner can be run on any web page and has a full team of experts ready to help extract information with custom instructions.

Key features of this plugin include:

Extract Tables and Lists: Easily extract information from lists and tables on websites.

One-Click Scraping: Extract data in seconds using ready-made instructions, and it can be downloaded as a CSV or xlsx file.

Extract Results from Multiple Pages: Automatically crawl pages and extract information from all pages.

Scrape and Open a List of URLs: Provide Data Miner with a set of URLs to scrape data from.

Execute Custom JavaScript: Use custom JavaScript to clean up scraped data, such as extracting emails.

Autofill forms: Automatically fills forms with the information it receives.

Downloading full web pages: Automatically downloading the full HTML of web pages from different sites. In fact, Data Miner scrapes URLs extract page information and load the page to your computer and download text, images, and every element from web pages.

Downloading images: Scrapes image links, extracts the image, and then sends the images to the computer. Once the images are saved, they can be uploaded to your site.

5.3.3.2 Google Translate Extension for Google Chrome Browser

Google Translate is a free multilingual translation service provided by Google for translating text. The service is available in various formats, including website, Android and IOS mobile applications, and an API that helps developers create extensions for browsers and software applications. Google Translate supports more than 100 languages at various levels and serves 500 million people daily as of May 2017.

Google Translate was launched in April 2006 as a US statistical translation service, using reports from the United Nations and the European Union Parliament to collect language data. Instead of translating directly, it first translates the text into English and then into the target language. During a translation, it looks for patterns across millions of documents to help decide on the best translation. Its accuracy has been criticized and mocked several times. In November 2016, Google announced that it was transforming Google Translate into a neural machine translation engine—Google Neural Machine Translation (GNMT)—that translates entire sentences at a time, not just fragments. Using this broader context, it helps identify the most appropriate translation and then rearranges and adjusts it to sound more like human speech with proper grammar. GNMT was initially enabled for just a few languages in 2016 and is gradually being rolled out to more languages. Google Translate can translate various forms of text and media, including text, speech, images, websites, or real-time video from one language to another. For some languages, Google Translate can pronounce the translated text, highlight different words and phrases in the source and target text, and act as a simple dictionary for word input. If the "Detect language" option is selected (the language of the text is not specified), text in an unknown language can be automatically detected. Users can save translations to a "phrasebook" for later use. For some languages, text can be entered via an on-screen keyboard, through handwriting recognition, or through speech recognition.

Google Translate is available for download as an extension for some web browsers. In February 2010, Google Translate was made available by default for the Google Chrome browser for optional automatic translation of websites. In this extension, by selecting the desired text on the website, we can see its translation, and it can also translate the entire page of a website instantly.

5.3.3.3 RapidMiner Studio Software

RapidMiner is a data science software platform developed by the company of the same name. It provides an integrated environment for data preparation, machine learning, deep learning, text mining, and predictive analytics. It is used for businesses and commercial applications, as well as for research, education, rapid prototyping, and software development, and supports all stages of the machine learning process, including data preparation, results visualization, model validation, and optimization. RapidMiner is developed in an open-source model. The RapidMiner Studio Free Edition, which is limited to 1 logical processor and 10,000 data rows, is available for free under the AGPL license. Commercial pricing starts at $2,500 and is available to developers.

This product was known as YALE (Yet Another Learning Environment) in its early years, in 2001, and was developed by Ralf Klinkenberg, Ingo Mierswa, and Simon

Fischer in the Artificial Intelligence Department of the Technical University of Dortmund. In 2006, it was developed by Ingo Mierswa and Ralf Klinkenberg, a company called Rapid-I in the same year. In 2007, the name of the software was changed from YALE to RapidMiner, and in 2013, the company Rapid-I was rebranded to RapidMiner.

This software has attracted the attention of many researchers and students in recent years due to its simplicity of operation, and on the other hand, the open-source development model of this product has also made programmers more willing to develop it. This software uses various algorithms for preparing and modeling information, and this wide range of options has led to many academic projects using this powerful product, at least for parts of their work. The graphical environment of this program is designed in a desirable and beautiful way, and users without programming knowledge can easily interact with different parts of the program. This tool provides users with the ability to connect to different programming languages such as Python and R. Even users of the Weka tool are no exception to this rule and can connect to this tool. RapidMiner's capabilities can be extended with additional plugins that are available through the RapidMiner Marketplace. RapidMiner Marketplace provides a platform for developers to create data analysis algorithms and publish them in the community. RapidMiner is written in the Java programming language, which provides a GUI for designing and executing workflows. These flows in RapidMiner are called "processes" and consist of several "Operators". Each operator in the process performs a single task, and the output of each operator forms the input of the next operator.

As mentioned above, this software has the ability to install various extensions and plugins. In this study, we use the text processing extension for text mining processes. This extension adds all the operators required for statistical analysis of texts and natural language processing (NLP) to the software. Texts can be loaded from many different data sources, transformed with a wide range of filtering techniques, and finally analyzed. The plugin supports multiple text formats, including plain text, HTML or PDF, as well as other data sources. It also provides standard filters for parsing, stemming, stop word filtering, and n-gram generation to provide everything for text preparation and analysis.

The plugin can be downloaded and installed via the RapidMiner Marketplace within the software or by using the link https://marketplace.rapidminer.com/UpdateServer/faces/product_details.xhtml?productId=rmx_text on the RapidMiner website – the operators of this plugin.

5.4 IMPLEMENTATION AND FINDINGS

In this section, we will carry out the steps mentioned in Section 5.3 in a practical and detailed manner. The research processes are also fully explained, and we will provide suggestions about their results.

5.4.1 Selection of the Products

The aim of this research is to analyze and examine the texts of customer opinions with the help of text mining and to increase customer loyalty by providing strategic solutions in the online store. The online store under study in this research is Digikala.

In the first step, we must select the products in question, about which the users' opinions are to be converted into knowledge. The segmentation of the products available on the website includes eight sections: digital goods, fashion and clothing, cosmetics and hygiene, books, culture and art, sports and travel, mother and child, and vehicles and industrial.

According to the studies conducted, the products were selected from the digital goods section. Some examples of these studies include:

- In the fashion and clothing section, most of the products in this section belonged to another site called Digistyle, and by reviewing that website, we found that its users had not given any comments, even though there was a specific panel for the comments section. In order to increase trust and attract more customers, in fact, to improve customer relationship management, the issue of encouraging customers to interact with the site should be addressed. Also, by further examining this section, we found that the best-selling and most commented products were smart watches and gadgets, which can be considered digital goods in a way.
- In the home and kitchen section, other than audio and video equipment and game consoles, other products received few comments.
- There were a few products in the automotive and industrial sector, of which a limited number had many comments, but even of those, very few were in the buyer category.

According to the above, the digital goods sector was among the most commented. Four products from the following sections: smartphones, tablets, external hard drives, and speakers were selected for exploration, which were among the bestsellers and included more comments than other similar products. Comments were extracted in order of their usefulness, not based on the date they were registered by users, which is listed in Table 5.1 of the selected products.

5.4.2 Extracting Comments from Selected Products

After selecting the desired products, we proceed to the stage of extracting data and customer comments about the four products mentioned. Using the Web Scraping technique explained in the previous section, using the Data Miner plugin, four instructions were written for the selected products.

The columns and features extracted from each comment include:

- Comment ID
- Author name
- Date
- Comment title
- Strengths
- Weakness
- Comment content

TABLE 5.1
Four Selected Products

Product Image	Section	Total Number of Comments	Number of Selected Comments
	Samsung Galaxy Note 8 SM-N950FD Dual SIM 64GB Mobile Phone Digital product > Mobile > Mobile phone	825	500
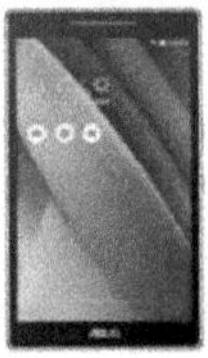	ASUS Zen Pad 8.0 Z380KNL 4G 16GB Tablet Digital product > Tablet and e-reader > Tablet	314	All
	Western Digital Elements External Hard Drive – 2TB Digital product > Computer and peripherals > Data storage devices > External hard drives	254	All
	JBL Go Portable Bluetooth Speaker Digital product > Computer and peripherals > Speaker	214	All

- Number of people who found the comment useful (yes)
- Number of people who found the comment not useful (no)
- Ratings for various product features, such as build quality, value for money, ease of use, design and appearance, etc., which vary depending on the number of columns for a specific product
- Buyer or not
- Customer recommendation to others if they are a buyer, which includes three states: agree, disagree, no idea (suggestion)

With the studies conducted on Persian language text mining solutions, the preprocessing of text data in RapidMiner software, especially in the stemming process, was fraught with problems, complexities, and limitations. As a result, translated data is used in this research.

Then, using the two extensions Data Miner and Google Translate, the written instructions are executed simultaneously, and the result of this execution is saved in the form of a table with the mentioned columns and rows that are the same as the comments in the .xlsx format in Excel software. Two data series were extracted from each product, one in Persian and the other with an English translation.

5.4.3 Initial Data Preparation and Cleaning in Excel

Before data preprocessing, which includes text mining processes, the data needs to be cleaned and edited to obtain more accurate results. This operation was performed in Excel software, which we will describe below.

Another column was added for the comment date; in fact, the solar date was recorded in one column and the Gregorian date in the other column. The editor add-on was used to convert the dates to each other.

In the comment title column, missing values were filled with frequently used texts from the same column, such as the product name, as shown in Figure 5.1. These missing values are more visible in the strengths and weaknesses columns. Also, in the buyer suggestions column, the cells where users were not buyers are also of this type, which are replaced with a dash (-).

In a questionnaire, the column names of each of the four products are also specified as product rating criteria, where users rated each product based on specific criteria from 1 to 5. As is clear, there is a Total Rate column in all products, which is a kind of average of the other scores.

A column called Total Rate was created to calculate the overall score of a product for each comment, which was calculated by averaging the scores of each feature and rounding it. To calculate the average, the formula =AVERAGE() was used in Excel, and then the average value was rounded to zero decimal places using the Round function in the KUTOOLS plugin.

The name of the buyer column was changed to Purchasing, which includes two states Buyer and Not buyer. Also, a column called user type was added, which has three types of values: buyer, commenter, and guest. And finally, a column called category was added for each table, which included the values (mobile, tablet, hard, speaker).

After editing the columns, we come to editing the text data, for which we used the editor software. At this stage, the following operations were performed:

- Converting colloquial texts to formal (you can't←you can't/for←for/ big←big/who/expensive←expensive)
- Spelling errors
- Correcting words that interfered with the translation
- Edit Markups
- Convert spaces to semi-spaces in two-part words and verbs (I go ← I go, Digikala ← Digikala, Amazing ← Amazing)

After performing the above operations, the translation process was performed again for better accuracy. The cleaned data of the JBL speaker product resulted from the initial preparation and cleaning stage.

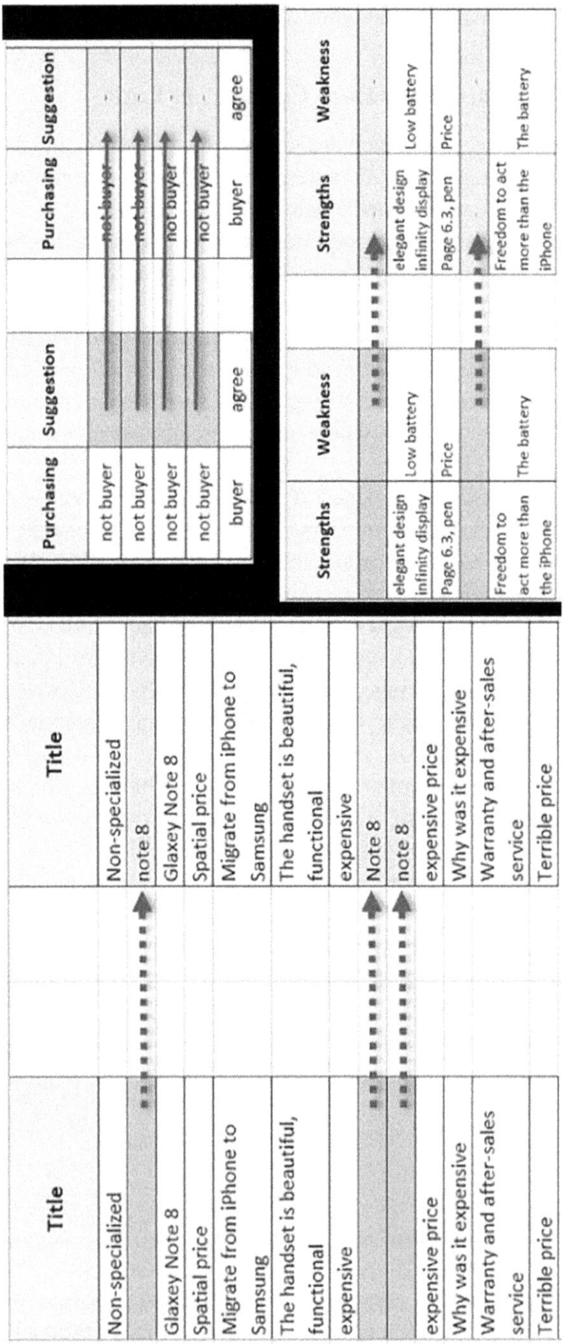

FIGURE 5.1 Replacing missing values.

5.4.4 RM Environment and Research Processes

After initial data preparation and cleaning, we enter the RapidMiner software environment, or RM for short. RM consists of two important design sections where processes are formed and the results that display the results of executing the processes. Data sets in RM are expressed as ExampleSet, and in the example tables, the rows and attributes are our columns.

In this research, eight processes are designed using RM. In the following, we will explain the details of each of these processes and examine their results.

5.4.4.1 Transferring Data and Generating the Required Datasets (F1 and F2)

After preparing and initially cleaning the tables, we transfer the data to the RapidMiner software to run the research processes. Our data consisted of four products, each of which consisted of two series, English and Persian, so at this stage, eight Excel files were added to the RM repository from the add data section. During the data import operation, we change the Comment ID column from regular to ID, as it has a unique value for each row, and ID columns do not affect the results in data mining calculations.

The other data that we need for the upcoming processes is a combination of data from the four products, which we need to create a process to merge these tables (F1).

In RM software, some operators include subprocessing; the first operator in the process is the Select Subprocess operator of this type (1). This operator can have several subprocesses, only one of which is executed. Here we have two subprocesses (2). In the parameters section of the Select Subprocess operator, the number of the desired subprocess is determined (if we enter parameter 1, our merged data will be English, and if we enter 2, our merged data will be Persian). The output of this data is entered into the next operator called Select Attributes, which, because in data integration, the columns of the tables must have the same number and type, we need to filter some columns, which is the responsibility of this operator (3). The columns that included scoring product attributes, except for total rate, will not be present in the output of this operator. After selecting the desired features, we reach the Append operator, which contains several input ports. Here, four ports are used for four products to merge two or more data sets, and the output consists of a merged data set (4). The Multiply operator does not make any changes to the input data, in fact, it only produces multiple copies of them. In this process, of the three copies used, one of them is connected to result, which displays the results, one uses the Store operator to store the merged data set in the RM software repository so that it can be used in subsequent processes, and the other is connected to the Write Excel operator, which saves the results as an Excel file in the path we give it (5). The result of this process contains two data sets, English and Persian. The results include 1,278 rows or examples, 14 columns with regular roles, and one column with Id role, and the values of the category column include 4 values (Hard, Mobile, Speaker, Tablet). In the missing section, it is clear that we do not have any missing data. The range column indicates the range of values for each attribute, and the name section specifies the names of those attributes, which are actually the names of the columns in our tables. The type column specifies the data type, with the attribute date being of type date, total rate, yes and no being of type integer, and the other attributes being polynomial or multi-valued.

In Figure 5.2, the scatter of the data obtained from F1 is displayed. The x-axis represents comment ID, the y-axis represents the Gregorian date, and the colors represent our four product categories. According to the dispersions, the newest comments are related to the mobile phone because the product is newer, and the oldest are related to the external hard drive. Of course, in the hard drive, the dispersion of comments is balanced across different dates, and considering that the product is not new, it also has new comments, which indicates that customers still welcome old products if they are of good quality. The dates of the comments on the speaker and tablet are also approximately proportional to each other.

The second process (F2), called generate test and train data, is created to slightly improve the performance of F8. This process uses sampling to generate test and training data for the prediction algorithm.

In part (1), the data of four products is placed, then with the Filter Example operator, we delete the comments that had a total rate of zero (2), because it means that no vote was registered for any of the product criteria, and the user in question did not rate it. Also, for mobile, we deleted 2, in addition to zero, because only one person had rated 2. In the next step, using the Numeric to Polynomial operator, we change the data type of the total rate from a number to a number of values (3). In part (4), using the Set Role operator, we label the variable that we are going to predict; in fact, we determine the role of the total rate, which is called the label of the labeled column values of the class. Then we come to the main operator in this process, the sample operator (5), which creates a sample of the data as training data. The relative sample parameter is set, meaning that a fraction of the total data is sampled. Then the balance data parameter is also selected to specify the proportion of samples in each class. For classes 3, 4, and 5, the ratio is 0.7, and for classes 1 and 2, due to their limited number, it is 0.5. The ratio 0.7 means that, in this process, 70 percent of the data with the specified class is assigned to the training data set, and using the subsequent operators, 30 percent of them will be considered as the test data set. Now, the two output ports of the sample operator, which are the exa (example) port of the training data and the ori (original) port of the total data, are entered as inputs into the subprocess operator (6). In the subprocess created using the Set Minus operator, the training data is subtracted from the total data, which forms our test data (7). Finally, for each product, two data sets are created, one for training and one for testing. After executing the process, we store these data sets in the RM repository as data needed for future processes.

5.4.4.2 Checking Active Users (F3)

The third process (F3), called Duplicate-Author, is created that aims to find users who have commented more than once and are considered active users to some extent.

The initial operator is Select Subprocess, which enters the merged Persian and English data into the next operator by selecting ports 1 and 2, respectively (1). This merged data is shown in the subprocess in part (2). The next operator is Filter Examples, which removes guest users from the data (3). In the parameters of this

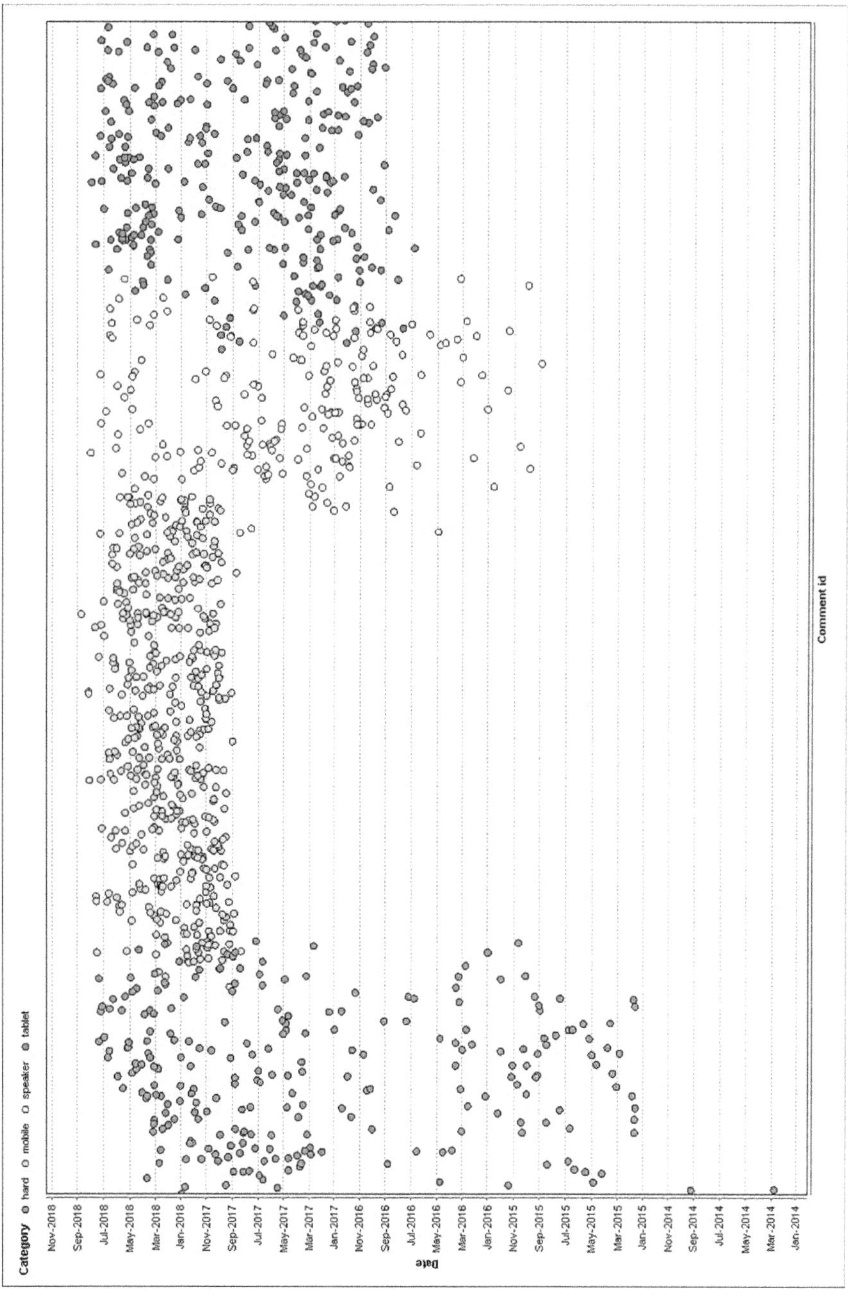

FIGURE 5.2 Data dispersion in merged comments (F1).

operator, we specify that if the Author column contains the words guest or guests, those records should not be present in the output by checking invert selection. Then, the Aggregate operator summarizes the data (4). In the parameters of this operator, we specify the count function for comment ID and group the data by Author. The output of this operator is a table containing two columns—Author and Count(comment ID)—which specifies the number of times each Author has been repeated. The exa port of this operator enters Filter Examples (2) which selects columns where count (comment ID) is greater than 2 (i.e., users who have commented more than once) (5). Since the output at this stage contains only two columns, it must be joined with the dataset from the ori output port of the Aggregate operator (6) where Author is specified as the key attributes parameter in the Join operator. In the Reorder Attribute operator, the order of the columns is rearranged using the ordering attributes parameter (7). Finally, the output enters the Multiply operator, and the results are stored.

In the Results and Charts sections, we see a scatter plot of the output data from F3. According to the shape of the x-axis, it indicates the number of times each user has commented, which includes four values, but for better clarity of the data, it has been enlarged and converted to decimal form. The y-axis also shows the Gregorian date, which is ascending from bottom to top. The colors in this chart also serve as our Authors.

The results of this process include 2 datasets: Persian and English. The output table of this process contains 157 rows. In fact, out of the total 1,278 comments, 175 were made by guest users, and among the other users, 67 people left more than one comment.

5.4.4.2.1 Analysis of Observations

The distribution of the number of buyers and non-buyers was regular in other products, except for mobile, where only one buyer had commented twice. Although the comments in the mobile section are more than in other sections in terms of both number and detail and accuracy of product reviews, it has fewer buyers, which is a sign of users' lack of confidence and trust in purchases of large and expensive items. There is a need to attract customer loyalty and trust in this type of product.

Some people had left comments for several different products. Many of the comments were left over a period of several months or even a year, including complete analysis and reviews. Some of the other comments were also written in a single day and sometimes contained similar themes or repeated texts. The presence of people in group 2 is a good sign that users continuously share their opinions during and after purchasing the product at different time intervals, which has a significant impact on users who intend to purchase. In a way, it attracts new customers.

In fact, these users, as well as users in group 1, are active users who need more attention and service in order to retain them and increase their loyalty. Users in group 3 actually have little impact on improving the business. This impact may be increased by giving them more awareness. Despite having a separate section called Q&A, some of the comments have expressed questions!

Some of the comments also express demands such as: product availability, availability of a specific color of the product, and special offers. Changing the website's

UI/UX to direct the user to the Q&A section or allowing the comment to be replied to in the same comments section might help improve this issue. By adding a section to express user demands, this weakness can be turned into a strength, which will have a huge impact on sales.

5.4.4.3 Analysis of User Types and Suggestions for Different Products (F4)

The fourth process (F4) named User type-category is created; its input is the merged data, and it has three different outputs. First, the merged Persian data is entered into the Multiply operator, which creates three copies of the data.

The first copy is the middle row operators, whose parameters are specified. First, the data is entered into the Select Attributes operator, which selects the comment ID, category, and user type columns for the output. Then, these columns are entered into the Aggregate operator, where a count function is created for the category, and grouping is done based on category and user type 3. The output includes a table with three columns: category, User type, and Count(category), which displays the number of different types of users (Commenter, Buyer, Guest) for each product. Then, using the Pivot operator, we convert the vertical table to a horizontal table to make the data easier to read. The output includes the columns: user type, Count(category)_hard, Count(category)_mobile, Count(category)_speaker, and Count(category)_tablet. Then, using the Generate Attributes operator, we calculate their percentage based on the total number of comments for each product. The formulas used are also specified in the parameters form. Then, using the Rename operator, we also change the names of the columns, and using the Reorder Attributes operator, we specify the order of the columns.

The results of the first copy include a table with 9 columns and 3 rows. The User Type column shows the type of users, and the highlighted columns indicate the percentage of each type of user for each product.

The results are drawn using the advanced chart in RM. The x-axis shows the user type, and the y-axis shows their percentage for each product. According to the chart, the speaker has the most buyers with 65%, and the mobile has the least buyers with 6%. The most guest users are also on the mobile and the least on the speaker. External hard drives and tablets are also almost equal, but the hard drive has more buyers and fewer guest users.

The second copy is the last row operators, whose parameters are specified. First, the data set is entered into the Select Attributes operator, which selects the comment ID, category, and suggestion columns for the output. Then we have the Filter Examples operator, which removes the rows whose suggestion column has a value of (-), which are actually considered non-buyers and therefore did not make a suggestion. Then, as before, the Aggregate operator is used, in which the count function is created for the category, and the grouping is based on category and suggestion. We also use the Pivot operator to rotate the table. In its output, the number of disagreements for mobile is indicated with a question mark, which is the missing data or missing value. We use the Replace Missing Value operator to convert its value to zero. Using Generate Attributes, we add a column called number of buyers, which is calculated by summing the total number of offers for each product. The Rename and Reorder Attributes operators also work as before.

The results of this copy include a table with 5 columns and 4 rows. Given that the ability to offer only exists for individual buyers, the number of buyers for each product is specified in the number of buyers column. In the other columns, we also see the number of types of offers for the four products.

The results are drawn using an advanced chart in RM. In all four products, the number of agreers was higher than disagreers, and there were also more people who did not have a suggestion than those who did. Even among mobile buyers, we did not have any disagreers, and there was one person who did not have a suggestion, which may be due to the low percentage of buyers in this product. After mobile, according to the number of buyers, the speaker had the most agreers and actually suggested the product, which encourages other users to buy this product. The tablet also had the most disagreers by a very small margin, who did not recommend the purchase of this product to others.

The third copy is the first-row operators. First, the dataset is entered into the Select Attributes operator, which selects the comment ID, category, suggestion, title, strengths, weaknesses, and content columns for the output. Then the Filter Examples operator separates the disagree suggestions for display in the output. The results of this copy include a table with 6 regular columns, 1 ID column, and 20 rows.

The comments of buyers who did not recommend the products to others are reviewed. By paying attention to, improving, and fixing the problems mentioned, these buyers will be satisfied in future purchases and will recommend the products to other users, which will attract more customers. To improve the problems mentioned in the table, more attention should be paid to the selection of future products, and these features should be checked by the quality control department. Problems during transportation should also be followed up in the postal and transportation department to prevent similar problems in the future.

5.4.4.4 Words and Adjectives Used in the Strengths and Weaknesses of Products (F5 and F6)

The goal in the fifth process (F5), which is called JJ-Weakness and Strengths, is to extract the attributes in the strengths and weaknesses for each product.

The first operator is Select Subprocess, which includes four subprocesses for the data of four products, which are processed in each selection of data of one of the products. The selected output enters the Subprocess operator; the output data in this operator is the same as the outputs of process 4. In the subprocess of this operator, the data first enters Select Attributes, which selects the comment ID, strength, and weakness columns. Then, with the help of the next operator Filter Examples, we delete the comments that have both strengths and weaknesses empty (have a value of (-)). The next operator is Nominal to Text, which changes the data type of strengths and weaknesses to text data. The data is then fed into the Multiply operator, and two copies are created, one for strengths and the other for weaknesses. First, we use the Select Attributes operator to select the strength column for the first copy and the weakness column for the second copy. The subsequent operators are similar in the two copies. This data is fed into the Process Documents from Data operator, which has two outputs, ExampleSet and wordlist, and actually performs the preprocessing of text data. This operator and the operators within its subprocessing are located in the text processing extension of the RM software. The first output port converts the text into a vector, which we will not use in this process, and the create word vector parameter is disabled in it. The second output port creates a list of words, which in the word column contains the names of the words created in the subprocess, and 2 other columns named document occurrences and total occurrences, which display the number of occurrences of the word in the document and the number of occurrences of the word in all documents, respectively. We also have another column named attribute name, which if the word vector is active, its values are the same as the values of the word column, otherwise it is empty. Now the data enters the subprocess of this operator. First, using the Tokenize operator, we divide the texts into smaller units called tokens, which are usually words. The next operator Transform Cases converts all letters to lowercase. Then, using the Filter Stopwords (English) operator, stopwords in English (in fact, words that have little semantic importance despite being frequently repeated in the text) are removed. In the Filter Tokens (by length) operator, tokens that contain less than two letters are removed from the text. Then, using the Generate n-Grams(terms) operator, we create sequences of up to 3 words from tokens, which are called trigrams. Finally, we have the Filter Tokens (by POS tags) operator, which uses the regular expression "jj.*" to separate all adjectives in the text and display them in the output. This operator is very time-consuming. For example, the mobile data, which is 500 comments, took about 12 minutes. The output port from this subprocess enters the Wordlist to Data operator, which converts the word list into a data set and changes the names of the two columns document occurrences and total occurrences to documents and total, respectively.

The sixth process (F6), named Weaknesses and Strengths, extracts the Persian words used in the strengths and weaknesses of the four products.

This process is similar in structure to the previous process, with only minor differences that we will discuss. The primary operator in this process is Subprocess, which contains the data we need. Due to the absence of the Filter Tokens (by POS tags) operator and the resulting time-consuming process, Select Attributes are not used, and the output data for all four products is created simultaneously. The Persian data for the products is also included in the subprocess of this operator. Then, the four outputs of the Subprocess operator are each fed into a Suprocess operator that is specific to a product. The structure of these four operators is exactly the same, and in fact, they only have different inputs. Another difference between this process and the previous process is its text data preprocessing part. The output data of the Process Documents from the Data operator is pruned; in other words, tokens with the specified number of occurrences are not displayed in the output. The parameters of this operator use the absolute pruning method, the lower limit of which is 2 and the upper limit of which is 9999. In fact, it means that tokens that are repeated less than two times in the documents are pruned. In the subprocessing of this operator, there is also the Tokenize operator first, and then, using the Filter Stopwords(dictionary) operator, we delete them through a text file containing Persian stopwords. In the parameters of this operator, we specify the file path and, since the content of the text file is Persian, we set the encoding to UTF-8. Then we have the Filter Tokens (by length) operator, and finally, in the Generate n-Grams(terms) operator of this process, 4-grams are used instead of trigrams. The wordlist output is also converted to an ExampleSet as before, using the Wordlist to Data operator.

The results of these outputs for the four products in process five are obtained in Table 5.2. There are two values of strength and weakness in the two columns above the table. For each product, there are two rows, one for the product attributes and the other showing bar graphs of those attributes. The bar graphs in the table depict the total values for the words. The x-axis shows the attributes, and the y-axis shows the total values.

5.4.4.5 Product Type and Score Prediction (F7 and F8)

The goal of the seventh process (F7), named k-NN Naive category, is to predict the category of the merged data using the k-NN and Naive Bayes algorithms. The parameters of this process.

The input to this process is the merged English data from F1. The data type of the title, strengths, weaknesses, and content columns in the merged data is changed to text using the Nominal to Text operator. The next operator is Select Attributes, which selects the title, strengths, weaknesses, content, comment ID, and category columns for its output. Then, using the Set Role operator, we determine the role of the column on which the classification and prediction are to be performed, that is, the category, as a label. Classification is used to assign a label to a set of data that has not yet been classified. The data is then assigned to predefined classes based on their characteristics. The classes in this process are hard, mobile, speaker, and tablet. In other words, classification is used to learn rules or build a model to predict the classification of new data. For classification, we need two datasets: a labeled training set used to build the model and an unlabeled test set (to achieve performance in classification, the test data is also labeled). We create these two datasets using the next operator, Split Data. This operator creates an arbitrary number of subsets of its input data according to the relative size specified in the parameters; in other words, it partitions the data. The number of outputs is based on the number of partitions specified. We set the sampling type parameter to shuffle sampling, which creates random subsets of the input data set. We also specify the partition parameter to two values, 0.3 and 0.7, which are the ratios of the data to be divided into two partitions. As a result, 70% of the data, consisting of 895 comments, is used for training and 30%, consisting of 383 comments, is used for testing. Then, the text data is preprocessed and weighted using the Process Documents from Data operator. The wordlist output in the training data must be connected to the wordlist input in the test data in order to perform prediction. This operator actually takes care of the word weighting operation and converts the text data into numbers so that they can be used in prediction algorithms. There are various weighting methods, the explanations of which were given in Section 5.3. We select the desired method from the parameters and vector creation section. In this process, we use term frequency, which is calculated by dividing the number of occurrences of a word in a document by the total number of words in that document. The tokens here are the words that are placed in the columns, and the documents in the rows and cells between them are the number of TFs. If the number of TFs of each token is greater, its weight is greater, and as a result, its importance increases. The first operator for preprocessing texts is Tokenize, then Transform Case is used, then stopwords are removed with Filter Stopword, and then we have Filter Tokens (by length), which removes tokens containing less than 2 letters. The

TABLE 5.2
Extracted Features in F5

<table>
<tr><th></th><th>Strength</th><th>Weakness</th></tr>
<tr><td>Hard</td><td>good | suitable | reasonable | beautiful | elegant | lightweight | proper | silent | authentic | easy | nice | appropriate | cheap | convenient | excellent | great | reliable | acceptable | affordable | attractive | capacity_reasonable | competitors_popular | continuous | core_beautiful_powerful | deceptivedefective | delicate | easy_run_good | effective | extremely_affordable | fair | flawless | good_quality_reliable | good_speed_proportional | idle | inexpensive | low_price_good | low_weight_structural | motionless | mute | popular | positive | powerful | proportional | quick | reputable | rigid | rigid_temperature_resistive | simple_stylish | slow | speed_proportional | strong | structural | tedious | unobtrusive</td><td>weak | low | sensitive | delicate | poor | fragile | vulnerable | absolutely_incomplete | anti_dirt_waterproof | bad | bodily_material | cheap | color_plastic | defective | bag_available | hot | impossible | incomplete | intermediate | intermediate_cable | lack_beautiful | loosening_usbnot_anxious | moderate | non_waterproof | old | poorly_impossible | read_variable | satisfactory_lack_beautiful | slightly_weak | slow | stubborn | thick | ugly | ultra_sensitive | vulnerable_reasonable</td></tr>
<tr><td>Mobile</td><td>great | dual | excellent | good | powerful | strong | beautiful | awesome | superb | functional | superior | unique | stunning | extremely_powerful | efficient | elegant | optical | slim | smart | accurate | attractive | easy | great_screen | innovative | sensitive | suitable | unlimited | amazing | big | brilliant | cool | dual_cameras_excellent | durable | ergonomic | excessive | extraordinary | flexible | free | full | pen_beautiful | pen_useful | professional | quality_camera_beautiful | quality_powerful | remarkable | sleek | unsurpassed | useful | absolute | abundant | acoustic | applicable | artificial | available | awesome_camera_super | awesome_pen_useful | awesome_screen_awesome | awesome_sensation_excellent | awesome_soundboard_good | bigger_benchmark_waterproof | bright | bright_screen | cheerful | comfortable | complete | connect_dex_stylish | density_pixel | design_ergonomic | designed_zoom_typical | device_superbly_functional | dex_stylish | dimensional | display_powerful | displayspen_giant_technical | dual_camera_optical | easy_use_comfortable | effective | enjoyable | excellent_pen_waterproof | excellent_</td><td>low | good | expensive | weak | inappropriate | bad | poor | great | big | low_internal | proper | thin | vulnerable | weaker | astonishing | bit_expensive | broke | card_thin_susceptible | careful | difficult | fragile | greedy | heavier_previous | inappropriate_location_priced | lcd_vulnerable | location_poor | minor | minor_mistakes_ignored | negative | old | pen_expensive | poor_battery_backup | poor_battery_priced | poor_facial | poorly_protected_weak | possibility_damage_poor | power_expensive | pretty_expensive | shorter | sim_cards_wired | slightly_weak | slim | slim_thin | slow | susceptible | thin_susceptible | tight | totally_vulnerable | unable | unfortunately_disappeared | unused | weak_physical | weaker_battery_previous | wrong</td></tr>
</table>

(*continued*)

TABLE 5.2 (Continued)
Extracted Features in F5

<table>
<tr><th></th><th>Strength</th><th>Weakness</th></tr>
<tr><td></td><td>speed_excessive | extensive | extreme | extremely_beautiful | extremely_durable | extremely_excellent | extremely_extraordinary | extremely_fast_dual | extremely_fast_powerful | extremely_flexible | extremely_superlative | extremely_waterproof | exynos_processor_tangible | facial_recognition_beautiful | fantastic | fascinating | fascinating_screen | fully_functional | fully_responsive | fully_responsive_size | gb_internal | gigabyte_rom_remarkable | good_display_powerful | good_screen_powerful | great_features_pen | great_quality_screen | great_screen_excellent | great_sense_touch | great_vibrant | happy | highly_efficient | highly_functional_handsfree | highly_powerful | infinite | iris_fingerprints_facial | loud | lovely | movie_speed_psychic | multi_layered | nice | optical_zoom_optical | original | pen_ultra_functional | pencil_photographic | performance_extraordinary | photographic | precise | psychic | quick | rear_camera_powerful | reasonable | responsive | responsive_size_suitable | rom_remarkable | rugged | selfie_spectacular | significant | sim_dual | simultaneous | smooth | soft | solid | spectacular | speed_psychic | strong_hardware_innovative | strongest | stunning_display_various | sturdy | super_awesome | super_functional | tall | tangible | technical | tiny | typical | unbelievable | unique_stylish | unmatched_display_super | unparalleled | various | versatile_functional | wonderful | world_features_suitable</td><td></td></tr>
<tr><td>Speaker</td><td>good | excellent | loud | clear | superior | beautiful | suitable | easy | great | strong | awesome | slim | clean | cool | durable | good_bass_excellent | quick | simple_stylish | superb | acceptable | adjustable | amazing | appropriate | authentic | awesome_sound_good | bluetooth_base_suitable | clear_sound_excellent | clear_sound_suitable | clear_voice_good | comprehensive | convenient | depending_size_good | desirable | dimensional | distinct | easy_carry_beautiful | excellent_material | fair | fair_prices_famous | famous | fast_charging_easy | fast_trouble_free | good_branded_authentic | good_charge_reasonable | good_quality_beautiful | indoor_</td><td>low | weak | expensive | bluetooth_weak | input_ram_simultaneous | late_charge | maximum_sound_monotonous | poor | slow | sound_monotonous | supporting_simultaneous | waterproof_supporting_simultaneous |</td></tr>
</table>

<table>
<tr><td></td><td>package_favorite | interesting | loud_noises_easy | loudness_volume_good | lovely | nice | nice_clear | notes_based_musical | perfect | proper | quality_good_solid | quick_easy | reasonable | reputable | separation_musical | sexy | simple_noise_excellent | solid | sound_base_desirable | sound_quality_reputable | sound_separation_musical | sound_suitable | sound_transparency_durable | transparent | unbelievable | unbelievable_size | unbelievable_size_good | unique | unparalleled | unparalleled_sound | unsurpassed | volume_beautiful | volume_volume_beautiful</td><td></td></tr>
<tr><td>Tablet</td><td>good | beautiful | excellent | strong | awesome | great | powerful | proper | nice | elegant | normal | affordable | appropriate | comfortable | modern | reasonable | suitable | superb | superior | acceptable | clear | compatible | great_tablet | perfect | relative | smooth | sturdy | attractive | bad | beauty_affordable | big_awesome | capable | card_compatible | classical | complete | convenient | conversational | cool | correct | delicate | dual | durability_quality_reasonable | durable | easy | efficiency_various | elegant_design_white | excellent_material | excellent_volume_conversational | extremely_fast_responsive | fairly_reasonable | fast_responsive | flexible | grateful | great_quality_build | great_screen | great_tablet_comfortable | hardware_acceptable | hardware_features_modest | hot | language_user_professional | light_reasonable | light_sensor_magnetic | loud | loud_speaker_powerful | loudspeaker_quality_excellent | magnetic | margin_application_powerful | modest | mute | nice_design_good | nice_tablet | normal_display_normal | original | otg_external | otg_screen_tasty | otgexcellent | personal | professional | regular | relatively_light_reasonable | responsive | robust | robust_battery_suitable | sensor_magnetic | sim_card_compatible | slim | slim_battery_awesome | soundboard_acceptable | thick | ultra_sharp | unique | user_professional | volume_conversational</td><td>low | weak | poor | slow | expensive | higher | normal | available_gb_available | disappointing | gb_available | hot | inappropriate | interchangeable | irreplaceable | lack_homogeneous | low_durability_cable | low_power_poor | low_ram_slow | nice | sleek | slim | slow_internal | strange | typical | unfulfilled_android | use_external | weak_camera_logical | weak_camera_suitable | widescreen | wrong</td></tr>
</table>

next operator is Stem(porter), which performs the stemming operation using Porter's algorithm. Finally, using the Generate N-grams operator, we used trigrams for the k-NN algorithm and bigrams for Naive Bayes (to prevent a large volume of data in the results). Because there are two algorithms in this process, the Select Subprocess operator is used, which includes 2 subprocesses, and by selecting each of them, the results related to the same algorithm are obtained. In subprocess 1, the k-NN operator is placed, where the value of K is 8, and the Euclidean distance is used. (This value has already been determined using the Optimize Parameters operator.) In subprocess 2, we also place the Naive Bayes operator. The next operator is Apply Model, which applies the model to the test data. This operator includes 2 input ports, model and unlabeled data, for training and test data, respectively. Finally, to evaluate the accuracy and performance of the prediction in this process, we use the Performance operator. We select two parameters, accuracy and classification error, for this operator. The value of classification error is the same as accuracy-1 or the classification error rate. The results obtained in this process for each algorithm include 3 outputs: model, predicted dataset, and algorithm performance.

5.4.5 Results of the k-NN Algorithm

The output of the k-NN algorithm model indicates that the learning model includes 895 comments and 50133 columns with K = 8 and classes hard, mobile, speaker, and tablet.

The output of the ExampleSet of this algorithm consists of 383 comments, which were our test data, 50133 columns, and 8 special columns. Of these 8 columns, one column is the comment ID with the role of ID, one column is the category column that shows the actual class of the comment, and the other is the prediction(category) column that shows the predicted classes, and we also have 4 confidence columns for each class. The confidence value actually determines the percentage of predicted confidence for belonging to each class, and for each row of the confidence column, the name of that class is placed in prediction(category). We filtered these results into 2 categories: correct and incorrect predictions, which were classified as 323 correct comments and 51 incorrect comments.

The output of the algorithm is obtained. The accuracy of this algorithm using the confusion matrix or classification matrix is 86.68%. Its error rate is 13.32%. The diameter specified in the matrix is the number of correct predictions for each class.

To better understand this matrix, the dispersion of the predicted data set is analyzed. The diameter specified in the matrix is the same as the ellipse in the diagram. 65 comments were in the hard class and were predicted to be hard, 3 comments from the hard class were classified in the mobile class, 4 comments in the speaker class, and 2 comments in the tablet class were misclassified, which can be clearly seen in the data scatter diagram.

5.4.6 Results of the Naive Bayes Algorithm

The output of the model in the Naive Bayes algorithm has three parts. The first part of the description shows that the output of the model has 23,832 columns, and the numbers generated for each class indicate the probability of the data being in that class. Given that the number of comments in the mobile class is more than the other three classes, this probability is also higher in it. The next part of the model output is the chart, which creates charts according to the specified attribute. In these graphs, the x-axis is the value, or in other words, the weight of the desired words, and the y-axis shows the density of that value in different classes, with the color corresponding to that class. Since our data was numerical, the value of $g(x,\mu,\sigma)$ or the density function was calculated in this graph (y-axis) according to Equation (5.8). The third part is the distribution table, which specifies the types of features that each of the attributes has. In this table, the mean parameters (mean= μ) and standard deviation (standard deviation = σ) of each word in different classes are calculated. For example, the column corresponding to the mobile class is arranged in ascending order, and the word note is placed in the highest row, so its number of occurrences in this class is also high. In the chart section, we also set attr to note. In the curves created, it is clear that the part where note is zero has the highest density for the three classes: hard, speaker, and tablet, and the green curve that represents the mobile class has densities in non-zero parts, indicating the presence of this word in the mobile class and its almost absence in the other three classes.

This density is more tangible in Figure 5.3, where the x-axis indicates the value of the TFs of note and the y-axis indicates the predicted classes. The colors also indicate the actual classes. In the part where TF is non-zero, data dispersion is observed only in the mobile category.

The output ExampleSet of this algorithm is the same as the previous algorithm, except that its number of columns is less, which is due to the use of bigrams instead of trigrams. The number of columns is 23832, a view of which is shown. Also, 313

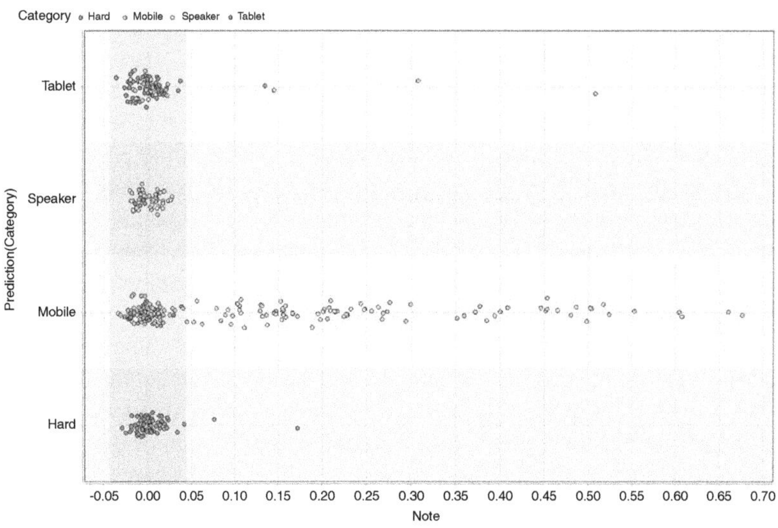

FIGURE 5.3 Distribution of the word "note" in the categories of the Naive Bayes algorithm (F7).

correct comments and 70 incorrect comments were classified. The accuracy of this algorithm using the classification matrix is 72.81%. Its error rate is also 28.18%.

The accuracy and errors of comment category prediction, as well as the mean precision, mean recall, and F-score metrics for the two algorithms k-NN and Naive Bayes are obtained. The mean recall and mean precision values were also obtained by averaging the class recall and class precision values in the classification matrix of these two algorithms. According to the results and performance of the classifications, the k-NN algorithm outperformed Naive Bayes and is a better method for classifying comments. Given the high accuracy for both algorithms, this method can be used to classify more products with a larger number of comments whose categories are not known and achieve good results. We conclude that by text mining the comments, their classification into different categories has been successful. Also, by using the chart in the Naive Bayes algorithm, it will be easy to find the density of the desired word types in different categories.

The eighth process (F8) called k-NN total rate predicts the overall score for each product using the k-NN algorithm. The input data in this process are the same as the outputs of F2.

The primary operators of this process are two subprocesses, one for preparing training data and the other for testing data. This operator includes four types of outputs, which are Hard, Mobile, speaker, and tablet products, respectively. Similar operations are performed in the subprocesses of these two operators, and only the initial data in them is different. This initial data is the training and testing dataset of the four products of F2 obtained through the sampling operation. Then, this data is entered into the Nominal to Text operator, which changes the data type of the title, strengths, weakness, and content columns from multiple values to text. Then, using the Select Attributes operator, the title, strengths, weakness, content, comment ID, and total rate columns are selected for output so that other columns do not affect the results of the algorithm and the modeling operation is performed according to the text data. In the next step, we have two Process Document from Data operators, which are test and train for preprocessing the text data. In this process, the TF-IDF criterion is used to weight the words. The calculation formulas for this method are stated in Section 5.3. The textual data preprocessing is the same as the previous process, except that the Generate N-grams operator uses 4-grams. The train data is fed into the k-NN operator to train the model as given in Figure 5.4. The optimal K value in this process is 16. The preprocessed test data and the data obtained from the k-NN operator are fed into the Apply Model operator to apply the classification. We also use the Performance operator to calculate the accuracy of the results. The output schema of the ExampleSets in this process. For hard, Mobile, Speaker, and Tablet, the number of examples is 68, 149, 65, and 95, respectively, and the number of columns or attributes is 10783, 41892, 8794, and 17034. The confidence columns that specify the probability of each comment's score include 5 columns. Only on mobile do we have three columns for confidence because there were no comments with scores of 1 and 2.

The classification matrix calculates the accuracy of this prediction for each. The accuracy and errors of comment prediction are also shown, as well as the mean precision, mean recall, and F-score measures of the comment scores for the four products. The mean precision measure for the external hard drive is higher than the others

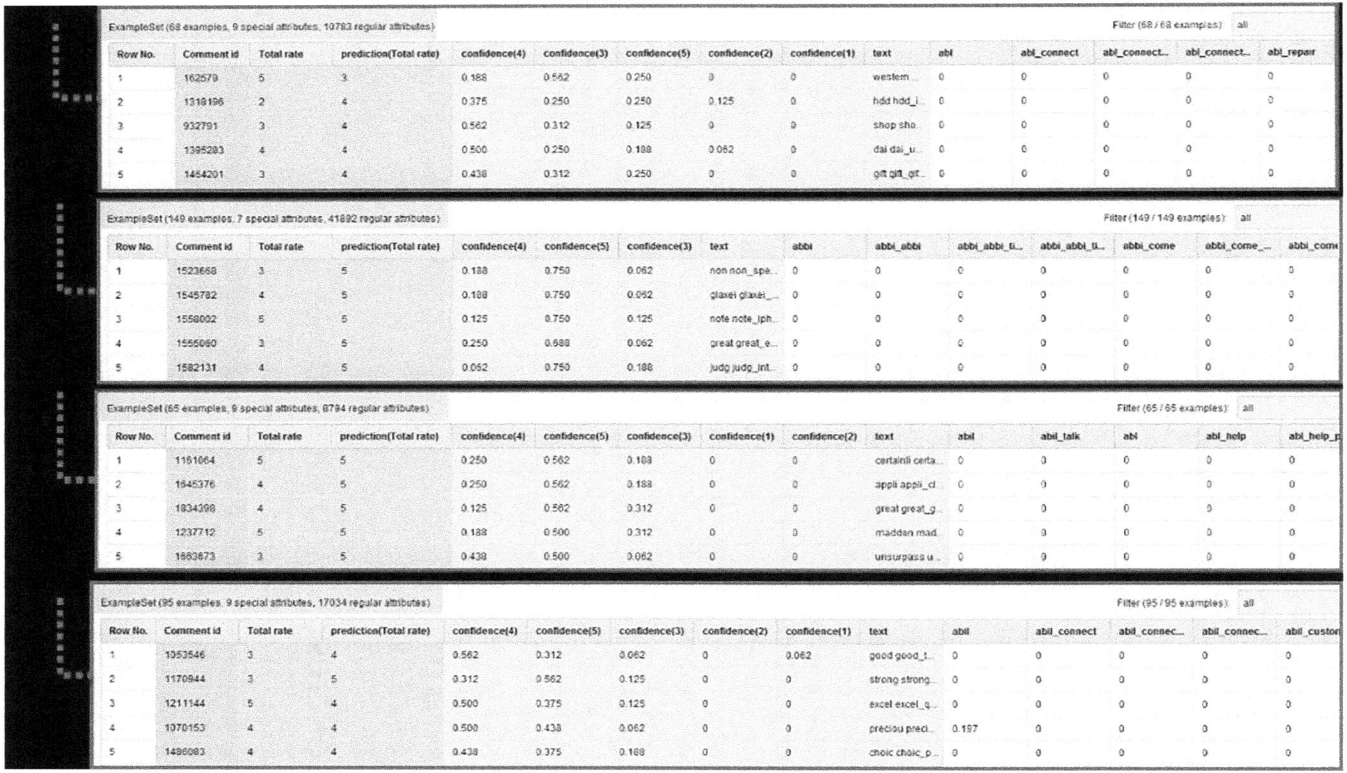

ExampleSet (68 examples, 9 special attributes, 10783 regular attributes) Filter (68 / 68 examples): all

Row No.	Comment id	Total rate	prediction(Total rate)	confidence(4)	confidence(3)	confidence(5)	confidence(2)	confidence(1)	text	abl	abl_connect	abl_connect...	abl_connect...	abl_repair
1	162579	5	3	0.188	0.562	0.250	0	0	western ...	0	0	0	0	0
2	1318196	2	4	0.375	0.250	0.250	0.125	0	hdd hdd_i...	0	0	0	0	0
3	932791	3	4	0.562	0.312	0.125	0	0	shop sho...	0	0	0	0	0
4	1395283	4	4	0.500	0.250	0.188	0.062	0	dai dai_u...	0	0	0	0	0
5	1454201	3	4	0.438	0.312	0.250	0	0	gift gift_gif...	0	0	0	0	0

ExampleSet (149 examples, 7 special attributes, 41892 regular attributes) Filter (149 / 149 examples): all

Row No.	Comment id	Total rate	prediction(Total rate)	confidence(4)	confidence(5)	confidence(3)	text	abbi	abbi_abbi	abbi_abbi_ti...	abbi_abbi_ti...	abbi_come	abbi_come_...	abbi_com
1	1523668	3	5	0.188	0.750	0.062	non non_spe...	0	0	0	0	0	0	0
2	1545782	4	5	0.188	0.750	0.062	glaxei glaxei_...	0	0	0	0	0	0	0
3	1558002	5	5	0.125	0.750	0.125	note note_iph...	0	0	0	0	0	0	0
4	1555060	3	5	0.250	0.688	0.062	great great_e...	0	0	0	0	0	0	0
5	1582131	4	5	0.062	0.750	0.188	judg judg_int...	0	0	0	0	0	0	0

ExampleSet (65 examples, 9 special attributes, 8794 regular attributes) Filter (65 / 65 examples): all

Row No.	Comment id	Total rate	prediction(Total rate)	confidence(4)	confidence(5)	confidence(3)	confidence(1)	confidence(2)	text	abl	abl_talk	abl	abl_help	abl_help_p
1	1161064	5	5	0.250	0.562	0.188	0	0	certainli certa...	0	0	0	0	0
2	1645376	4	5	0.250	0.562	0.188	0	0	appli appli_cl...	0	0	0	0	0
3	1834398	4	5	0.125	0.562	0.312	0	0	great great_g...	0	0	0	0	0
4	1237712	5	5	0.188	0.500	0.312	0	0	maddan mad...	0	0	0	0	0
5	1663673	3	5	0.438	0.500	0.062	0	0	unsurpass u...	0	0	0	0	0

ExampleSet (95 examples, 9 special attributes, 17034 regular attributes) Filter (95 / 95 examples): all

Row No.	Comment id	Total rate	prediction(Total rate)	confidence(4)	confidence(5)	confidence(3)	confidence(2)	confidence(1)	text	abil	abil_connect	abil_connec...	abil_connec...	abil_custor
1	1053546	3	4	0.562	0.312	0.062	0	0.062	good good_t...	0	0	0	0	0
2	1170944	3	5	0.312	0.562	0.125	0	0	strong strong...	0	0	0	0	0
3	1211144	5	4	0.500	0.375	0.125	0	0	excel excel_q...	0	0	0	0	0
4	1070153	4	4	0.500	0.438	0.062	0	0	preciou preci...	0.197	0	0	0	0
5	1486083	4	4	0.438	0.375	0.188	0	0	choic choic_p...	0	0	0	0	0

FIGURE 5.4 Output dataset of the k-NN algorithm (F8).

and is a measure of its accuracy, and scores of 3 and 4 were well predicted in it. The highest performance in various measures, except for the mean precision, is for overpredicting the scores belonging to mobile. The reason for the higher percentage of mobile compared to other products is that, firstly, the number of comments was higher, and secondly, the number of its classes was 3, which resulted in a better prediction accuracy value, which in fact, most of the comments were assigned to one of the three classes, which caused the other two classes to not be classified correctly.

5.5 DISCUSSIONS

In this process, considering the percentage of accuracy of the predictions, these scores need to be improved. Given these results, a number of users do not give the correct opinion. The accuracy of the given scores is very effective in the output. To be more accurate, it is better to increase the scoring intervals, assuming that instead of 1 to 5, users have the ability to score between 1 and 100. Also, if different products, such as the mobile category, have more comments, we will have a more accurate prediction, so there is a need to encourage users to comment and rate their experiences with products correctly so that the comments increase and, in turn, we have a more accurate classification.

The results of the two initial processes (1 and 2) produced inputs for other processes. In process (3), active users who had commented on the products several times were extracted, and then we analyzed the text of their comments, which were more important than those of other users. The results of process (4) were displayed visually in graphs, and the types of users and their suggestions were examined according to the graphs. Information about the number of guests, buyers, and non-buyer users for different products, buyers' suggestions, and paying attention to dissatisfied comments play an important role in increasing profits, loyalty, and customer satisfaction. In the next two processes (5 and 6), we examined the words and attributes used in the strengths and weaknesses of the products, the results of which will be of great importance for selecting the store's next products. In the last two processes (7 and 8), we also used data mining algorithms to create categories based on the type of product.

5.6 CONCLUSIONS AND SUGGESTIONS

As mentioned, online shopping has higher advantages compared to physical shopping in stores, which is why the level of competition among online stores is very high. Given that customers play a key role in these businesses, the need for customer relationship management is felt more than ever. In order to retain existing customers and attract new customers, user opinions and comments require attention, which is why in this study we analyzed the comments of users of the Digikala website. Initially, different product categories were examined to select the desired product for analysis. The results of the observations were that digital goods were better received for providing opinions. To improve this situation, by advertising, gaining trust, and providing better services, the number of comments on non-digital products can be increased, which will increase the number of buyers by reading the comments if the services are of high quality. Following the above, four products were selected from the digital goods section: external hard drive, mobile phone, speaker, and tablet, then extracted, translated, and stored in Excel tables. An analysis of the comments was created in the form of processes in the RM software. The achievements of the results of the processes are as follows:

- We merged the product data and visually reviewed the comments left for each product based on date.
- We considered users who had left multiple comments as active users and reviewed their comments. By paying attention to the comments of these users and meeting their needs, these users will become loyal customers who recommend purchasing from this store to others, and in turn, the sales volume will increase.
- Studies were also conducted on different types of users and their suggestions. To convert guest commenters into site users and then buyers, we need a visual diagram of the number of this type of user for different products. One of the features of the feedback section is for product buyers to recommend or not recommend the product to others. We presented these two needs in the form of graphs for easier understanding by business analysts by creating a process. Other suggestions for more optimal analysis of different types of users include creating a panel as a profile that has access to their other information, such as age, gender, favorite products, field of study, and other characteristics of a customer, and can benefit from them in processes to improve the business.
- On the website, users have the opportunity to comment on the product in two formats: strengths and weaknesses separately. We extracted the most frequent adjectives and words for each product in the form of functional words and graphs by performing text mining on these sections. These results will have a significant impact on the smarter selection of future products in the store.
- We used two algorithms k-NN and Naive Bayes in data mining to classify texts that were converted into numerical data by text mining preprocessing and TF and TF-IDF weighting. In simpler terms, we predicted product categories. The accuracy of the k-NN and Naive Bayes algorithms was 67.86% and 72.81%, respectively. We also calculated the average recall and precision classes and the F-score value for the two algorithms. Given the high accuracy obtained in this classification, it can be used for higher volume data that do not have a specific category.
- We created training and test data through sampling to predict the comment scores for different products using the k-NN algorithm. We calculated the accuracy, average recall, and precision classes, and the F-score for each product. The results of this classification indicated that users were not rating accurately, and some users had not even entered a rating. To improve the accuracy of this prediction, we need to increase the details of the rating intervals and also encourage users to leave more comments.

This study can be used as a support for management decisions in website design and product introduction, as well as the use of suppliers in line with customer interests.

User comments are considered very valuable information for improving online businesses and meeting customer needs. Therefore, conducting various researches to convert this data into knowledge for the online business community is essential. The following are suggestions for future research in this area:

- Using Python for Persian text mining of comments using the Digest library and using stemming techniques in Persian
- Text mining and sentiment analysis of tweets containing the hashtag "DigiKala"
- Comparing reviews of similar products between Iranian online stores and the Amazon website
- Comparing customer satisfaction metrics between multiple online stores
- Using databases with larger data volumes for better accuracy in text mining
- Using a larger number of products in the same category as well as analyzing non-digital products
- Predicting future customer purchases using association rules
- Analyzing sentiment in different product categories and social networks such as Instagram

6 Social Network Analysis for Product Recommender System in Online Retailing

6.1 INTRODUCTION

A recommender system or recommendation system is a subset of an information filtering system that seeks to predict the "score" or "priority" that a user gives to an item. Recommender systems are used in various fields, including movies, music, news, books, research articles, search queries, social tags, and products.

Recommender systems usually produce a list of recommendations in one of two ways – collaborative filtering or content-based filtering (also known as the personality-based approach). Collaborative filtering approaches create a model of a user's past behavior (previously purchased or selected items or numerical ratings for those items) as well as similar decisions of other users. This model is then used to predict items or rank items that the user prefers. Content-based filtering methods use a set of discrete features of an item to recommend more items that have similar properties. The above approaches are often combined, which is called a hybrid approach.

Recommendations play an important role in our daily lives. Recommender systems automatically recommend items to the user that they may be interested in. Recent studies show that social network information can be used to improve the accuracy of recommendations (Fazlollahtabar, 2020).

Various techniques of recommender systems have been proposed since the mid-1990s. Researchers and managers believe that recommender systems provide many opportunities for business, government, education, and other fields. Online business transactions and e-commerce success depend heavily on the effective design of the product recommendation mechanism. Traditional recommender systems, especially collaborative filtering (CF) recommender systems, have been studied by many researchers in the past decade. However, they ignore the social relationships between users. In fact, this relationship can improve the accuracy of the recommendation. In recent years, the study of social information-based recommender systems has become an active research topic (Bagheri et al., 2018).

Recommender systems are software tools and techniques that provide recommendations for items that a particular user is likely to be interested in. Recommender systems are personal information agents that provide recommendations. Recommendations for items

DOI: 10.1201/9781003740148-6

that the user is likely to use. The recommendations are intended to provide interesting elements to users. Recommender systems can be developed using various techniques and algorithms. A recommender system customizes its responses to a specific user. These systems are designed to serve as an information agent.

People who share a workspace usually share information. Such information is generally obtained through collaboration, when individuals help each other to achieve collective or individual goals. Recommender systems have been developed to help bridge the gap between information collection and analysis by filtering all available information and presenting relevant items to the user. A recommender system helps to increase the capacity and efficiency of this process. Typically, there are two different types of recommender systems: (1) Collaborative filtering and (2) Content-based methods. The first method is based on recommending items to individuals who have similar opinions or tastes to the target user. The second method is a recommendation method based on the item's attributes and measuring the similarity between the attributes and the target user's preferences.

With the rapid growth of Internet websites and social networks, various services are provided on these platforms. For example, Facebook focuses on social activities, and Twitter and Plurk (also called microblogs) both focus on user interaction through short messages. Millions of users enjoy these websites, which have marketing capabilities. Understanding and knowing users can help companies to increase the accuracy and efficiency of their target market. Therefore, in this study, we intend to propose a social recommender system based on microblog data. This social recommender system is built based on the messages and social structure of target users. The similarity of discovered features of users and products is calculated as the nature of the recommender engine (Tajdin et al., 2007).

Users have a lot of information at their fingertips that they are unable to process. Hence, the need to develop new tools is felt. Recommender systems have been designed to meet this need, and many techniques have been used to approach this problem in different ways. To provide relevant recommendations, these systems use large datasets and do not use the social network. Creating a recommender system that takes into account the users' social network is another way to deal with this problem. The aim of this project is to use the theory of six degrees of separation among social network users to improve existing recommender systems.

The following assumptions are used in this thesis:

- Social networks are used as the main source of discovering people's tastes and preferences.
- Users' preferences, tastes, and preferences change over time.
- The concept of click stream is used to examine knowledge in social networks in line with the product.

6.2 BACKGROUND

Recommender systems help users identify items of interest. These recommendations are generally made in two ways: by calculating the similarity between items and suggesting items related to those the user has expressed interest in, or by calculating the similarity between users in the system and recommending items that are of similar interest to the users. The second method is known as collaborative filtering. When people make decisions, they usually rely on the recommendations of friends and acquaintances. Although collaborative filtering – the most popular recommendation method – uses similar neighbors to make recommendations, it does not distinguish friends in a neighborhood from strangers who have similar tastes.

Collaborative filtering (CF) is the process of filtering or evaluating items through the opinions of others. CF technology brings together opinions from large connected communities on the web, supporting the filtering of significant amounts of data. CF is a well-known method in recommender systems and uses the relationships between users and recommended items with the active user and with respect to the ratings of their neighbors. This method suffers from the problem of data scarcity, where users only rate a small set of items. Thus, the calculation of similarity between users is uncertain, thus reducing the accuracy of CF algorithms.

Currently, social networking websites easily extract social network information; studying the use of social network information in creating more accurate recommendations is likely to have positive results. With the increasing number of available social networking services in which individuals can invest their time and effort, we see this investment being made by the user, which includes providing personal information, creating and sharing content, establishing social connections with other users, and interacting with other users and content in a myriad of possible and visible ways such as Facebook, YouTube, LinkedIn, Twitter, etc. (Fazlollahtabar, 2010).

The main motivation of the idea is that the value of personal information related to an individual or some digital content (text, audio, video, etc.) increases with the knowledge of other entities in those relationships. With the expansion of e-commerce, both individuals and organizations are producing and consuming information online. With the availability of a lot of product information on websites, many customers are confused by this huge amount of information about the items they want to buy. Recommender systems (RS) are a solution to this problem.

Many websites such as YouTube, eBay, and Amazon have provided their own versions of recommender systems. However, issues such as lack of information, changing data, changing user settings, and unpredictable events are among the problems of recommender systems in providing a suggestion or recommendation. Also, one of the important problems of recommender systems can be considered the problem of cold start and changes in user tastes and preferences, which double the difficulties in achieving an accurate recommendation (Marzouni et al., 2018).

Proposing a new recommender system that includes user profile parameters obtained from social networking websites could be a solution in this regard. The proposed model could exploit a collaborative filtering algorithm focusing on user preferences obtained from social networks. The continuous and diverse growth of social networks has led users to interact with them more. With these changes, what used to manage only a limited social connection is now used to exchange ideas and opinions and create a need for new features (Fazlollahtabar and Ahmadiahangar, 2024).

Users have a lot of information at their fingertips that they are unable to process. Hence, the need to develop new tools is felt. Recommender systems have been designed to meet this need, and many techniques have been used different approaches to solve this problem. To provide relevant recommendations, these systems use extensive data sets and do not use the social network. Creating a recommender system that also considers the users' social network is another way to implement a technique that will ultimately improve the accuracy of recommendations.

Social network data, represented by FOAF terms, is one of the most common types of semantic web data. Using a weighted average recursive search, one can consider two people in the network and arrive at a recommendation or suggestion about how trustworthy the other person is based on the paths that take them through the network and the validity of those paths (Fazlollahtabar, 2021a).

Recommender systems are an important part of the information and e-commerce ecosystem. They are a powerful way to enable users to filter through big data and product spaces. Almost two decades of research on collaborative filtering have resulted in a diverse set of algorithms and a rich set of tools for evaluating their performance. Research in this area is moving toward a more complete understanding of how technology advances in specific domains.

The diverse aspects demonstrated by different algorithms of recommender systems show that recommendation is not a one-dimensional problem. Specific tasks, information needs, and the target domain pose unique challenges for recommender systems. The design and evaluation of recommenders should be based on the tasks at hand. Effective deployment should begin with a careful analysis of potential users and their goals. Based on this analysis, system designers have many options for selecting an algorithm and embedding it in the user experience of the surrounding environment (Alikhani and Fazlollahtabar, 2014).

There is an incentive to create recommendations for content based on social networking communities. For example, a user may receive a list of items of interest based on content viewed by others in the user's social networks. Recommendations may be based more on content information, such as polls, ratings, tags, attributes, and the like, from various sources internal and external to the user's social networks. Content items are given a weight that corresponds to a certain level of relevance or interest to a user. Using the weight, a list of recommended items is sorted or filtered. In one or more configurations, the weight changes based on the age of the content elements. For example, the relevance, importance, or interest of a news report may decrease as the news becomes older and older (Fazlollahtabar, 2012c).

Many researchers have focused their efforts on creating collaborative recommender systems. It has been proven that the use of collaboration in such a system improves performance. But what is not clear is how this collaboration is carried out and, more importantly, how collaboration should be carried out in order to optimize the exchange of information. The collaborative relationships in recommender systems can be introduced as a social network. Several measures can be proposed to analyze collaboration based on social network analysis. Once these measures are explained, they can be used to evaluate a specific example of participation in a real-world proposal system.

6.3 PROBLEM DEFINITION AND MODELING

Recommender systems are techniques used to provide item recommendations to users. Formally, a recommendation problem can be defined by the usability of the function "rec". This function is defined as the function that predicts, for a given user u, the item i from a set of items I among the set of users U. Rec is the function U*I R. R is in the interval [1, 0] and is the usability score of the recommended item. Rec refers to the capabilities and abilities of the item to meet the needs of the users. In this regard, the task of the recommendation system is to define this usability score for a hypothetical user and item. In general, the data and information available to the recommender system (RS) can be very diverse.

Items are the objects and targets of the recommendation (offering recommendations). They are defined by their complexity and usability. The complexity of an item is expressed by its structure, representation, and dependencies on other items. Naturally, a recommender system (RS) recommends a specific type of item (movie, music, etc.).

Users of a recommender system (RS) are very diverse in terms of their interests and goals. The recommender system (RS) identifies and specifies recommendations using information related to the users. This information can be organized in different ways, and the recommendation techniques used define and specify the user information of the model.

Transactions define the interactions between users and the recommender system (RS). Transactions are the data resulting from interactions between humans and the recommender system. The types of information used for recommendation techniques, such as the item selected by the user and the description of the research context, are very different. The transaction can also include explicit feedback from the user. This process is naturally called evaluation, such as rating an item. Various recommendation techniques have been proposed in previous articles and research. Including the content-based approach of recommender systems, the participatory approach of recommender systems, the statistical approach of recommender systems, the knowledge-based or profit-based approach of recommender systems, the collective approach of recommender systems, and the hybrid approach of recommender systems.

The concept of criticism (evaluation) is very important in our framework and perspective. This concept is based on the idea that users specify their demands and requirements as goals that are not met by conventional item recommendations. The evaluation-based recommender system interprets the users' unconstrained preferences to identify the congruence and alignment of item features with the users' preferences. The major steps of the criticism-based recommender system include the following:

Item recommendation: This step selects a set of items *r* to be presented to the user. At the beginning of the evaluation cycle, the selected items are determined based on the user's search *q* (similarities between the requirements and candidate items).

Item review: In this step, the user reviews the proposed item in order to accept the recommendation or select an evaluation, which is the start of a new evaluation cycle. If a criticism is selected, only those items that meet the criteria defined in the evaluation are considered and analyzed further (this process reduces the candidate item set).

In this sense, our Intelligent Recommender System (IRS) offers and develops ideas beyond traditional recommender systems, as it tries to understand users, discover their interests, and the like through knowledge creation, reasoning, etc. During these processes, we use the concepts of "evaluation unit" and "criticism components".

Nowadays, recommender systems are widely used to support collaborative tasks. However, it is important to consider the information of each user of the system about the proposed topic. We describe the purpose of user information to improve the Business Process Cooperative Editor (BPCE) recommender system – a collaborative business process modeling tool. We use the concept of information vectors, which was developed in previous research on collaborative orientation, to bring user information into the recommender system. A simulation was designed to evaluate the performance of the new recommender system, which we named the editor. In the following, we present information vectors and how they are used for the editor.

6.3.1 Knowledge Vectors

Since each person has a different level of knowledge about a topic, it is natural to weigh the user's suggestions against their knowledge of the topic. Therefore, we need a way to relate the user's evaluation of an item to their knowledge of the content of that item. Our first challenge is to calculate how much the user knows about the topic. We use knowledge vectors to solve this problem. Basically, the knowledge vector (KV) is the user's information and knowledge, which is represented to the user as a sum of periodic vectors (term vectors) for the suggested items.

Since the user's suggestion will be used to represent their knowledge of the topic, another important issue is the accuracy and precision of the user's evaluation of the item. This process is needed since the user can suggest many items about a topic, and according to the model presented above, their level of knowledge about the topic increases. However, their suggestions may not be accurate and correct, and as a result, they will not be able to correctly add periodic vectors to their information vector. This problem can be solved by using a reputation system. In addition, this process may also provide a solution to their situation when there are bad users in the reputation-participatory network systems.

Regarding the second problem, what we want is to avoid unknowingly adding periodic vectors to the user's information vector, and strengthening their knowledge about a topic when they do not have the necessary information, for example, they do not know much about the topic. This can also affect the final ranking of the item. To solve this problem, we add an accuracy factor called Evaluation Accuracy (EA) to the information vector formula. Evaluation Accuracy is calculated by averaging the evaluations of each page. The logic behind this is to consider the mean as the accurate and correct evaluation of the page, since most users make their evaluations based on their knowledge of the topic and their environment, and the environment does not know how others evaluate it. Therefore, we penalize users who evaluate a page outside the mean of their evaluation. In addition, we reduce the penalty according to the available evaluations. For example, if the dispersion is low, we consider the mean to be a legitimate expression of the "truth" compared to when the dispersion is high. This process is done by considering the standard deviation in the denominator of the correct factor.

6.3.2 Modeling

In this section, we define the important components of our intelligent recommender system. The difference with the information-based recommendation approach is that there are four important elements in an intelligent recommender system, which are mentioned in the following section:

- Information evaluation mechanism, which is based on algorithmic learning.
- Explicit information modeling, which represents all the information necessities and requirements in order to provide a recommendation.
- Reasoning mechanism to infer information from stored data.
- Evaluation system based on the automatic inference capability of the intelligent recommender system (IRS).

Our intelligent recommender system not only provides items based on specific information related to how the user's needs and preferences are met by the item's characteristics and features, like the information-based recommendation approach, but our system also uses item ranking, such as the collaborative approach, to discover the user's desired aspects, interests, and features related to the items (criticism system), and infers the item ranking. In this regard, this process combines the ideas of traditional recommender systems based on existing similarities to automatically infer the user's needs for items with the concept of inferential criticism to automatically infer user goals that are not included in the conventionally considered items. The intelligent recommender system extracts all the knowledge that has been automatically obtained (by the training mechanism) and appropriately applies reasoning mechanisms to infer the user's need for the item in such a way that the user's goals are met with common items. Basically, two aspects need to be defined in our intelligent recommender system. These two aspects include the structure (scheme) and the information model.

In this area, the important component is the semantic information model that stores the different types of information used for the proposed items. This information must be updated because it is needed for the learning mechanisms. Finally, in order to extract the information, a reasoning mechanism must be used that is responsible for suggesting items by using all the available information. Now, the following section lists the different components of the structure:

6.3.2.1 Information Modeling

An important aspect of defining a paradigm is the information representation. There are many paradigms – ontologies (natures), fuzzy rules, concept maps, etc. The important points for choosing one are the capabilities of representing all the available information and the possibility of defining reasoning mechanisms with them. In general, our proposed model requires different types of information, such as user model, context model, domain model, recommended item model, and other information related to user-item behavior that stores ratings, criticisms, etc. (which we call the critical model). The Intelligent Recommender System Framework section describes the different types of information in our Intelligent Recommender System in detail.

6.3.2.1.1 Information Acquisition

This stage is defined to learn common situations. There are many approaches (supervised and unsupervised), but the important point is to define an approach that allows for the discovery of all available information at a given moment. In general, different machine-learning techniques can be potentially used according to the search context. For example, online information and real-time information. The information source is very diverse and can be structured (e.g., transactional databases, or an organization's data repository), semi-structured (e.g., xml files), or unstructured (e.g., GPS tracking data, audio streaming systems, etc.) data that represents information related to users, search context, etc. Information is obtained through learning mechanisms based on data mining, semantic meaning (web mining, context mining, ontological mining), among other techniques. The learning mechanism used depends on the information model and the data source. For example, if the information model is an ontology and the data source is a web, we can use semantic meaning techniques to extract information. In addition, in this case, the science function for data extraction, cleaning, transformation, and reduction is very important before applying data mining learning techniques.

6.3.2.1.2 Reasoning Mechanisms

According to the choice of model, there are specific reasoning mechanisms that can be used. The important point is that these mechanisms should allow for inference. If the user's goals are not met by the common items, all available information should be considered. There are three important reasoning mechanisms that can be used, including analysis, comparison, and inference. Each can be used for different things like analyzing the reason for a recommendation, predicting which item might be of interest to the user, which means that it allows for different types of reasoning: finding a "relaxation break" or "agreement" (e.g., can the user's needs be met?), finding a "diagnosis" (e.g., why is a certain item recommended?), performing a "diversification" and "refund-debug" (e.g., is it incompatible with the user's needs?), and the like. During recommendation processes, the idea of defining reasoning explanations related to various aspects of the desired outcome is generated using the available information. Traditional recommender systems only address some of these needs.

6.3.2.1.3 Critical (Evaluation) System

An automatic system for inferring user preferences without asking users about them, and using only the information stored in the recommender system. In each cycle of the proposal cycle, there is a reasoning stage where aspects and issues are inferred for acceptance or criticism. These cycles continue until a proposal is made. Learning mechanisms provide the necessary information, such as a feedback mechanism, to infer the user's preferences. They obtain the necessary inferred information to infer the item ratings and build a set of proposal cycles. If a new cycle is targeted, then only items that meet the criteria defined in the evaluations are given more attention (reducing the set of candidate items, in order to reduce the search space: candidate items). In general, these processes continue until the reasoning mechanism determines that the user can accept the proposal and has explored all possibilities, or the proposal cycles end.

6.3.3 Information Model in Intelligent Recommender System

The important part of our intelligent recommender system is information. In an intelligent recommender system, all available information should be extracted, and all advantages in different domains (such as information retrieval and data mining) should be used to extract information. In the old and conventional way, a recommender system estimates the similarities between the item characteristics and the user's preferences, or the ranking of items that have not been viewed by the user. The intelligent recommender system extracts information to infer the ranking of items, to infer the user's preferences, and to align the item characteristics with the user's preferences. For this purpose, the system defines different types of information:

6.3.3.1 Detailed User Profile

This includes opinions, evaluations, and natural information about the user's preferences, as well as their personal information (age, gender, occupation, and education). In this approach, we propose to develop new information about the opinions, criticisms, and relationships with other users (e.g., friendship groups, etc.). The user profile model, along with the rest of the semantic models of the intelligent recommender system, should infer the preferences and needs of the users.

In this model, information about the users is defined. This model should allow answering questions such as the following: Who is the person? What personal information is available? How is he/she performing? What projects are planned? What tasks and activities have been performed? What skills does he/she have?

Some of this information is obtained by asking users questions, but some of this information is obtained through learning approaches. For example, friend networks can be obtained using social network analysis, or ideological orientation can be used in semantic mining mechanisms.

6.3.3.2 Detailed Item Profile

This profile describes a complete description of each item based on four dimensions: (i) General product description (name, branches that produce the items, etc.); (ii) Functional information about the item (its function, etc.); (iii) Structural information about the item (its components, relationships between them, dependencies, etc.); and finally (iv) Practical information about the item (how to use them, etc.). Some of this information can be learned or inferred from stored information.

In this model, we define a set of characteristics that represent an item. There are different types that can describe an item.

- General or generic features and characteristics, such as functional information, structural information, descriptive information.
- Specific or individual features and characteristics, which can be the level of use, rating, or usability.

These features can be represented by concepts or abbreviations that can describe their characteristics. These concepts are listed in the following section:

- Intrinsic or static (self)
- Dynamic (changeable features)

In the present study, we have characterized each item by four dimensions:

- Descriptive: information about its intrinsic and innate characteristics. Answer the question: What is it?
- Structural: the elements that make up the item (entity, features, processes), the relationships and constraints between them, etc. This part reflects the unchanging structure. Answer the question: What are its components?
- Functional: the specific functions of the item. Answer the question: What does it do or what does it enable?
- Functional: how is it usable? How does it interact or integrate with other systems? Answer the question: Is it usable or reusable?

Table 6.1 lists the various characteristics that describe an item in our system. (The first letter next to the item indicates whether it is static (I) or dynamic (D), and the second letter indicates whether it is descriptive (D), structural (S), functional (F), or operational (O).)

Here, some attributes are defined as specific or individual, but in fact they belong to critical information (evaluation), because their information is based on

TABLE 6.1
Item Profile in Intelligent Recommender System

Overall and General Properties	Internal and Subjective Properties
- Item ID (I, D)	- Usage level (D, O)
- Name (I, D)	- Utility (D, O)
- Item type (I, D)	- Emphasis/attribute (D, O)
- Description (I, D)	- Type of problems it can be used for (D, O)
- Localization (I, D)	- Reuse (D, O)
- Item author (I, D)	- Extensibility (D, O)
- Date details (I, D)	- Interoperability (D, O)
- Producer (I, D)	
- Dimensions (I, D)	
- Version (I, D)	
- Format (I, D)	
- Components (I, S)	
- Relationship between components (I, S)	
- Imposition (I, S)	
- Technical equipment (I, S)	
- Purpose (I, F)	
- Equipment coverage (I, F)	

the user's judgment and opinion. In our case, this judgment is inferred by using reasoning mechanisms beyond the stored information and the user's transactional information.

6.3.4 Evaluation Information

This section contains information that needs to be discovered based on the recommender system transaction (relationships between users and items). A transaction may describe the context of the recommendation, may imply the selection of items, and may include feedback generated by the user, among other things. Naturally, this information describes behavior-based information and may be discovered using machine-learning techniques to obtain desired patterns. This information naturally represents the interests of users. In our intelligent recommender system, this information needs to be trained or discovered.

In our case, the similarities between the item features and characteristics and the user's preferences, or the estimation of the item's ranking, result from the inference processes of the information model used (e.g., if the information model is ontological, then we can employ an ontological reasoning).

In general, in our intelligent recommender system, the available information at a given moment should be used without reducing its performance. This process depends on the domain and user scenario request, and only a part of the previous information may be available; the intelligent recommender system should continue to provide recommendations. In this regard, the definitions of different types of information are very important and significant; however, our intelligent recommender system is very powerful in working with the available information. The important difference with the information-based recommender system is that this system requires information engineering, and the intelligent recommender system uses training mechanisms from different sources, production databases, social media, etc. Our system can extract various machine-learning techniques and semantic mining in order to generate the required information. For example, if the information model is ontology, then we can use the integration and adjustment of ontologies (two types of semantic mining techniques) to achieve the model. In this regard, our intelligent recommender system avoids the information acquisition problems of information-based recommender systems. In addition, there are no ramp-up and cold-start problems, because the intelligent recommender system can make inferences about users and items even when sufficient information has not been collected. Its recommendations do not depend only on user ratings (as in a collaborative recommender system) or on collecting information related to a specific user or item (as in a concept-based recommender system). Because this system infers information.

Domain and context information: It is very important to know the domain in which the items will be used, the context in which people will make decisions, etc. The information domain is all the information that explains a hypothetical situation. The information domain is the information of a regular area, a human activity, etc. This type of information is not conventionally considered, or the recommender system must modify its structure in a conventional way for use in a specific context. Here, we propose an explicit model of these aspects.

Two additional pieces of information, in addition to what has been said, are included in the following section to achieve the semantic information of our intelligent recommender system:

- Context information: In this case, it is necessary to collect all the information specific to the context in which the recommendation is made.
- Information domain: This is the valuable information used in a hypothetical human activity domain, in a specific setting, etc. Naturally, the expert uses and develops his domain information. In the intelligent recommender system, this refers to the specific domain that will be used.

6.3.4.1 Evaluation Pattern

User behavior is modeled to determine his/her criticism, preferences, opinions, and tendencies. The pattern represents opinions and criticisms. Specifically, the user's opinions (criticisms and evaluations) about the suggested items are stored as a result of inference processes, training processes, computation (estimation), or querying of the user's inferences. This pattern is oriented for use during recommendation processes (suggestions) and the display of discovered preferences (this is the product criticism space). In this regard, the intelligent recommender system can recommend items that meet the conventional and common evaluations, which are similar to the previous recommendations. Our system infers the user's preferences and tasks through the built-in recommendation cycle.

6.3.5 Proposed Algorithm

Creation and Construction of Information Vector: In the following section, we list the formulas used to calculate user information, starting with the definition of some variables.

- EA_{ki} – the accuracy of user k's evaluation of item i.
- e_{ki} – the evaluation of item i by user k. Its value ranges from 0 to a maximum of mx.
- μ_i – the mean of item i before evaluation.
- mx – the highest evaluation an item receives.
- σ_i – the standard deviation of the item from the evaluation.
- TVi – the periodic vector of item i.
- KVk – the information vector of user k.

Evaluation Accuracy:

$$EA_{ki} = \begin{cases} 1 - \dfrac{|e_{ki} - \mu_i|}{mx \times \sigma_i}, & \sigma_i > 1 \\ 1 - \dfrac{|e_{ki} - \mu_i|}{mx}, & \sigma_i \leq 1 \end{cases} \tag{6.1}$$

The maximum value of the factor is 1, which is obtained when the user evaluates the page with a value exactly similar to the average score of the page.

In Formula (6.1), the numerator indicates how far the user's rating is from the mean, and the denominator indicates how spread out the ratings are for each page. The largest value of the numerator is mx, since the ratings range between 0 and the maximum mx, and the average value is between 0 and the maximum mx. This process occurs when the new rating is very different from the mean. In addition, the smallest value of the numerator is 0, which is observed when the new rating is exactly the same as the mean. When the largest value of the numerator is reached, for example, when the difference between the page rating and the average rating is close to the maximum, the Evaluation Accuracy (EA) approaches 0, preventing the periodic vector (TV) from being considered for calculating the user's information, since the user's rating of the item is very different from what is called the "correct" score. On the other hand, even if the hypothetical evaluation is very different from the mean, but the standard deviation is large, the user will receive the periodic vector (TV) almost completely on his information vector calculations, because we know that in this case the mean cannot easily guarantee "accuracy and precision".

Information Vector:

$$KV_k = \sum_{0}^{n} TV_i \times EA_{ki} \tag{6.2}$$

The information vector is calculated by summing the periodic vectors (TV) of all the items evaluated by the user, multiplied by their respective precision factor. Since the information vector is a multidimensional vector, the higher the dimension value, the more information the user has about the topic related to the dimension.

6.3.5.1 Using Information Vectors in Recommendations

With this path of computing user information, when an item is recommended to a user, the evaluation of each item will be calculated through the information of the user who made the evaluation. We calculated the final score of each item using the user evaluators measured through the similarities between the user information vector and the periodic item vector.

It is worth noting that our recommender system cannot avoid the uncertain start problem. The user information vector may be wrong from the beginning if it consists of a small number of recommendations. When more recommendations are generated, the information vectors become more valid and achieve the efficiency of the recommender systems. In addition, when the number of recommendations for each item is small, the mean and standard deviation are also incorrect, resulting in the information vectors of the users who recommended the item being suppressed. In short, recommender systems require a certain number of recommendations for each item, and active and effective users are more likely to be successful.

In the following section, we list the formulas used to calculate the final score of each item, starting with the definition of some variables.

- KF_{ki} – user k's information factor for item i.
- KV_k – user k's information vector.
- Tvi – periodic vector of item i.
- Si – total score of item i.
- e_{ki} – evaluation of item i by user k. Its value ranges from 0 to maximum.

Information Factor:

$$KF_{ki} = \frac{KV_k \cdot TV_i}{\| KV_k \| \| TV_i \|} \tag{6.3}$$

The information factor is the similarity between the user k's information vector and the periodic vector of item i. This factor is used to measure how much the user's evaluation was considered in the ranking of the item. As in the similarity comparison, if the cosine of the angle formed by the information vector and the term vector is 0 (zero), the user's evaluation will not be considered, since the user has no knowledge of the topic defined by the item. This can be considered as the user's credibility for the topic evaluation, or the degree of our confidence in the user's evaluation of the topic. Item Score:

$$S_i = \frac{\sum_{k=0}^{n} e_{ki} \times KF_{ki}}{n} \tag{6.4}$$

The final score of the item is used as its recommendation and is calculated by averaging the user ratings, weighted by the information factor related to their subject.

6.3.5.2 Enhancing the Business Process Collaborative Editor (BPCE) with Information Vectors

The Business Process Collaborative Editor (BPCE) is a P2P tool that allows designers to create new patterns and models based on existing ones and share their patterns freely. In its ranking algorithm, only the model scores are considered in gaining the designer's credibility. Basically, user information is not considered.

We integrated the Business Process Collaborative Editor (BPCE) with information vectors because the ranking chart did not take into account the designer's information, and as a result, there was a problem: a designer with a lot of credibility, regardless of their expertise or specific past work, could create models for any topic. The models could be considered good and appropriate only because of the author's credibility, regardless of whether they had information on the topic in question.

So, to avoid this problem, we integrated information vectors with the Business Process Collaborative Editor (BPCE) to include the designer's information related to a topic in the final model score. To use information vectors in the Business Process Collaborative Editor (BPCE), the formulas are changed to use the information factor along with the user's credibility. The formulas for calculating the designer's credibility have not been modified.

In the next section, we list the original formulas (a) and the modified formulas (b), starting with the definition of some variables.

- j – designer
- m and k – templates
- PImj – initial points of model m created and developed by designer j
- PMm – total points of model m
- PSm – selected points of model m
- PUm – used points of model m
- Rj – designer j's reputation
- PSjmk – points for selection (0 if no selection, 1 otherwise) by designer j, when model m is built from model k
- PUjmk – points used: when not used = 0, low use = 1, medium use = 2, and high use = 3, by designer j, when model m is built from model k

Initial Model Scores:

$$Pl_{mj} = \frac{\sum_x (PM_k \times \frac{pu_{jmk}}{3})}{|x|} + e^{(4 \times R_j)} \tag{6.5a}$$

The initial score factor evaluates a model based on the validity of the models used in its construction. This process is done considering that if a model is more advanced than other models, it must have some validity over previous models. This can be referred to as the model DNA.

The first operation of the formula, which is in the fraction part, is related to the model DNA. If the evaluated model receives the maximum evaluation (Pujmk = 3), the new model receives its full model score. The initial scores of all evaluated models are summed, and the fraction represents the arithmetic average of these scores.

The second operation, the power part, gives more points to the model if the model designer has a lot of validity. This process allows for rapid growth of the model score. We chose number 4 in the power because we want to reward the best designers, and we know that 0.3 is a good score for a designer, considering that we have more than 5 designers. In this case, the power increases quickly. In Formula (6.b5) that follows, we modify the formula according to the user's information about the item's subject.

$$Pl_{mj} = \frac{\sum_x (PM_k \times \frac{pu_{jmk}}{3})}{|x|} + e^{(4 \times R_j \times KF_{jm})} \tag{6.5b}$$

In this case, we used the information factor (KF) to measure the power effect in the formula. Users who have information about the subject of the item are increasingly rewarded according to their experience and expertise in the subject, and this process is consistent with power.

Model Selection Points:

$$PS_m = \sum_j \sum_k (R_j \times ps_{jkm}) \tag{6.6}$$

When creating new models, the selection points factor is expressed as the points accumulated by the model due to its selection by other designers. A model receives a reward at least for being desirable.

$$PS_m = \sum_j \sum_k (KF_{jk} \times R_j \times ps_{jkm}) \tag{6.7}$$

In addition, this reward is measured by the information factor of the design that selected the model. The principle is: the designer's experience and expertise tend to select better models rather than incorrect ones, and as a result, models receive more points when they are selected by designers who are experts in the subject.
Model Usage Points:

$$PU_m = \sum_j \sum_k (KF_{jk} \times R_j \times pu_{jkm}) \tag{6.8}$$

The usage points factor represents the points that models receive from other designers when they are evaluated for their usability in building new models. A model is rewarded for its actual use in building a new model. Also, as seen in Formula (6.9), with the information vector, the reward will be measured by the information factor of the designer who chooses it as the basis for his new model.

$$PU_m = \sum_j \sum_k (KF_{jk} \times R_j \times pu_{jkm}) \tag{6.9}$$

6.4 IMPLEMENTATION AND FINDINGS

We evaluated the use of information vectors for model ranking in the Business Process Collaborative Editor (BPCE) through simulation. We simulated the behavior of the Business Process Collaborative Editor (BPCE) twice using simulation techniques, once without information vectors, such as using the original ranking graph, and again with information vectors, in order to answer the following questions. The question to be answered is: Was the model that achieved the highest score and highest quality created and developed by an expert?

6.4.1 Scenario (Research Story)

In this simulation, we considered an organizational modeling activity that has ten different business branches and needs to model Procurement and Acquisition (P & A) processes. Each branch has a unique model. In addition, the organization has five business units with designers who are knowledgeable about the procurement and utilization processes, and five other business units do not have designers who are knowledgeable about the procurement and utilization processes.

6.4.2 Simulation Processes

We started the simulation by considering that each designer builds his or her own model of the procurement and utilization (P&A) processes. Then, each designer randomly decides to build a new type of model. When the designer builds a new type of model, he or she runs the build process with the old model and searches the repository to retrieve some other models that can help improve the existing model. When the new type of model is created, it overwrites the previous model, and the designer evaluates the imported models.

We noticed that when the new type of model is built, it inherits the quality of the previous type, but with modifications according to the designer's actual credentials. In other words, we considered that a good designer has a higher chance of improving their model, but may sometimes be demoted, while a poor designer has a lower chance of improving and correcting their model. Below we show the details of the setup and creation of the simulation described in Figure 6.1.

Input variables:

- i – Number of simulation iterations
- wi – Initial model weight
- ws – Model selection weight
- wu – Model usability weight
- m – Number of models in the repository before the simulation starts
- mo – Number of designers in each organization

Each designer has a quality factor that indicates how the designer performs, e.g., good, average, or poor. This factor is used to simulate the designer's quality. Similarly, each model has a quality value that is derived from the designer. Although the suggested number of groups is 4 or 5, we chose only 3 groups, good, average, and poor, for the simplicity of the simulation.

The values chosen for the model quality value are: good, if the quality value is greater than 80, average, if between 40 and 79, and bad if less than 39. We chose these values considering that the probability of a model being good is less than that of being average or bad, because we believe that it is not easy to create and develop very good models in the real world.

The third step, "Evaluation of imported models", is related to the number of points each imported model receives. To decide how many points are given to imported models, we considered that the points received by each model are also related to its quality, and therefore should also be related to the designer's level of awareness of the model-building processes.

We defined Qm as the quality value of the model, i.e., whether the model is good, average, or poor. The probability that a model will receive a score of 0, 1, 2, or 3 is

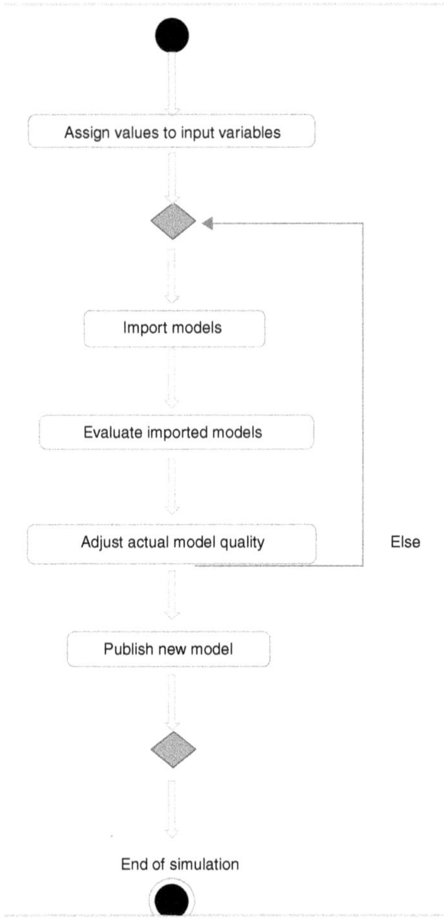

FIGURE 6.1 Simulation flowchart.

related to its Qm and is defined based on a Bernoulli distribution with the probability of success p as follows:

$$p = \begin{cases} \min p, & \text{if } Q_m = \max Q_m \ (a) \\ \max p, & \text{if } Q_m = \max Q_m \ (b) \\ 1 - \dfrac{Q_m - \min Q_m}{(\max Q_m - \min Q_m)}, & \text{else} \qquad (c) \end{cases} \tag{6.10}$$

where maxQm is the maximum value of Qm, minQm is the minimum value of Qm, minp is the minimum value of part (c) of p, and maxp is the maximum value of part (c) of p.

If a model has good quality, say 80 ≥ Qm, it will receive 2 or 3 points. Therefore, the probability p of a model receiving 2 points is defined as:

$$p = \begin{cases} 0.05, & \text{if } Q_m \geq 100 \\ 0.95, & \text{if } Q_m = 80 \\ 1 - \dfrac{Q_m - 80}{20}, & \text{else} \end{cases} \tag{6.11}$$

These values are chosen so that if the quality value of the model (Qm) is close to the upper limit of the scores of a typical model (40–79), the probability that the model will receive 2 points increases. If the quality value of the model is greater than 100, for example, when the model quality is high, the probability of receiving 3 points is higher. The actual p-value is chosen arbitrarily, according to our understanding of the chances of an expert making a rigorous assessment.

If the model has moderate quality, say 40 ≤ Qm ≤ 8, it will receive 1 or 2 points. Thus, the probability p of a model receiving 1 point is defined as:

$$p = \begin{cases} 0.025, & \text{if } Q_m = 79 \\ 0.975, & \text{if } Q_m = 40 \\ 1 - \dfrac{Q_m - 40}{40}, & \text{else} \end{cases} \tag{6.12}$$

These values are chosen so that if the quality value of the model (Qm) is close to the upper limit of the scores of a weak model (less than 39), the probability that the model will receive 1 point increases. If Qm is close to 80, for example, the quality of the model is high, and the probability that it will receive 2 points increases.

If the model has poor quality, say 1 ≤ Qm ≤ 40, it will receive 0 or 1 point. Therefore, the probability p of a model receiving a score of 1 is given by:

$$p = \begin{cases} 0.975, & \text{if } Q_m = 39 \\ 0.025, & \text{if } Q_m \leq 1 \\ 1 - \dfrac{39 - Q_m}{40}, & \text{else} \end{cases} \tag{6.13}$$

These values are chosen so that if the quality value of the model (Qm) is close to the upper limit of the scores of a weak model (close to 39), the probability of the model receiving a score of 1 increases. If Qm is close to 0, the probability of receiving a score of 0 is higher.

The above formula is applied according to the Bernoulli distribution with the probability of success p; for example, if the prior distribution is used and p is defined. If there is no success, the evaluation is carried out according to the uniform distribution (0, 3). In this case, we note that if the designer has a lot of credibility, they tend to evaluate the model according to a certain criterion and rule.

In the simulation without information vectors, the distribution is as follows:

$$p = \begin{cases} 0.9, & \text{if } \mathrm{R}_i > 0.15 \\ 0.6, & \text{if } 0.10 < R_i \leq 0.15 \\ 0.3, & \text{if } R_i \leq 0.10 \end{cases} \tag{6.14}$$

In the simulation with information vectors, the distribution is as follows:

$$p = \begin{cases} 0.9, & \text{if } \mathrm{R}_i \times KF_{im} > 0.15 \\ 0.6, & \text{if } 0.10 < R_i \times 0\,KF_{im} \leq 0.15 \\ 0.3, & \text{if } R_i \times 0\,KF_{im} \leq 0.10 \end{cases} \tag{6.15}$$

The next step is related to the quality of the newly built model. We examined the quality of the model to determine if it increased, decreased, or remained unchanged. If the quality of the model increased, its quality value increased by a factor of 5. If the quality decreased, its quality value decreased by a factor of 5. The rate of increase or decrease of the quality value was chosen experimentally and empirically. Other studies can be investigated with different settings.

We used experimental distributions with different values according to the designer quality factor:

Good designer: p(increase) = 0.5, p(unchanged) = 0.4, p(decreased) = 0.1.

Average designer: p(increase) = 0.2, p(unchanged) = 0.6, p(decreased) = 0.2.

Poor designer: p(increase) = 0.1, p(unchanged) = 0.4, p(decreased) = 0.5.

The final step of "create new model" is when we create the new model and start a new interaction.

The distributions used in this simulation have experimental and empirical values and are based on our observations from the real world. They represent the research where our recommender system has been evaluated.

6.4.3 Simulation Execution

In this section, we present the simulation execution parameters. Regarding the first simulation stage, the following values are set for each execution section. Different from the Business Process Collaborative Editor (BPCE) simulation intended in the research, we performed the simulation once with the best values of wu, wi, and ws found in this simulation.

- We performed i-90 interactions, considering that a designer creates 10 types of similar business processes:

wi – 0.1; ws – 0.3; wu – 0.6; m – 10; mo – 10.

For the designer quality factor, we selected two designers as good designers, two designers as poor designers, and six other designers as average designers. Since it is necessary that the sum of the designer credits is equal to 1 and different from the previous simulation, we have ten designers. We defined the initial values of the designers as follows:

- 0.15 – good designer credit
- 0.10 – average designer credit
- 0.05 – poor designer credit

To answer the proposed question, we ran the simulation processes 100 times and counted the number of times the model had the best quality in the first position, and the number of models built by an expert. We ran the simulation with the original ranking graph and with the new ranking graph as information vectors.

6.4.4 Simulation Results

In this section, we present the results obtained from the simulation. The results of the original ranking graph and the results of the new ranking graph are obtained. By analyzing the graph, we can conclude that without information vectors, most models are not created by an expert. Therefore, the chance of the best model being a "fake" person is high. Also, by analyzing the graph, we can conclude that with information vectors, most of the best models are created by an expert, increasing the quality of the recommender system. Therefore, we can conclude that the best models are the result of the activity of an expert in the modeling process, with information vectors, and consequently, increasing the efficiency of the recommender system. The percentage of best models created by an expert and without information vectors, is as follows:

Models created by an expert = 42%; Models created by a non-expert = 58%

The percentage of best models created by an expert and with information vectors, is as follows:

Models created by an expert = 65%; Models created by a non-expert = 35%

6.5 CONCLUSIONS AND DISCUSSIONS

In order to create and construct an information vector that helps us in designing a recommender system, we need to obtain the accuracy of user k's evaluation on item i. Using the proposed algorithm, we were able to prevent the inclusion of evaluations that have a maximum difference from the average evaluation and not consider a place for such evaluations in the periodic vector. Thus, the accuracy of the evaluation increases by eliminating evaluations that are contrary to the average evaluation and also placing evaluations that are in line with the average evaluation. On the other hand, even if the hypothetical evaluation is very different from the average, but the standard deviation is large, the user receives the periodic vector (TV) almost completely based on his information vector calculations, because we know that in this case the average cannot easily guarantee "accuracy and correctness". As a result, there is no need to worry about losing the evaluations of users who have evaluated items with a large standard deviation.

The information vector is obtained by collecting the periodic vector and combining it with the Evaluation Accuracy. As a result, with the information vector obtained and the periodic vector in hand, it is time to calculate the information factor. The information factor is the similarity between the user's information vector k and the periodic vector of item i. This factor is used to measure how much the user's evaluation has been considered in the ranking of the item. This can be considered as the user's credibility for the evaluation of the subject, or the degree of our confidence in the user's evaluation of the subject. Now, the item score should be calculated, which is obtained by averaging the user's evaluations and contributing the information factor. By obtaining the item score and the above-mentioned items, the conditions for modeling recommendations and suggestions are provided.

Today, there is a lot of information on the Internet, which is growing rapidly. Users' needs for specific information consumption and their inability to process all the information have led to the rapid development of information retrieval technologies. Search engines effectively perform the search task and make a large amount of information available to a wide range of users. However, most search engines provide the user with too much information to process. This problem is solved by the recommendation system. The main difference between recommendation system algorithms and data search algorithms is that they provide answers without an explicit request from the user, while search algorithms respond to a specific user request. Leading companies – search engine developers (Yahoo, Google) – have noticed that the future of the World Wide Web lies in recommendation systems. Search is still very important, but something more useful needs to be added. Personalization is a key feature that can be an advantage for these companies, providing recommendations based on the user's interests. The above approaches to building recommendation systems apply to first- and second-generation recommender systems. Third-generation recommender systems are currently at the beginning of their development. The main innovation of such systems is their orientation toward semantic models of representing and using knowledge, in particular, knowledge about the user's personal profile.

Currently, the most natural and advanced way of drawing semantic categories, which is commonly used in decision-making processes, is ontology. For this reason, in modern and future systems, which are called third-generation recommender systems, ontology is considered a general framework for presenting various and diverse types of knowledge. Examples of such knowledge are knowledge of the user's personal profile, the context of the decision, the user's emotional state when making a decision.

7 New Product Development and Multi-Factor Analysis

7.1 INTRODUCTION

In the current era, the advancement of science and technology has caused extensive changes in the world, and this has caused companies and manufacturing plants to try to improve in various sectors by utilizing these technologies more effectively. In this thesis, we also intend to first examine what effects technological advancement has had on manufacturing industries and how to best benefit from this technology in the future.

A product goes through several different processes from the time it is just an idea until it reaches the customer. In this thesis, we want to examine this chain and see how it can change in today's world. New product development is a business process for developing new products for a company. New product development includes all activities from developing an idea or a concept for a product to realizing a product through the stages of production and introduction to enter the market. After the new product development stages are completed, it is time to present the product to the market (Fazlollahtabar, 2008).

In today's world of constant connectivity, the customer journey from product discovery to final purchase often involves multiple purchase channels. It is not uncommon for shoppers to browse products in stores and then place their order online, or to research products extensively online before completing their purchase in a physical store. Consumers have become sophisticated enough to optimize their shopping experience by comprehensively examining all possible options across all possible channels. As a result, retailers are under great pressure to integrate the best of the physical and digital worlds at every stage of the customer experience (Tajdin et al., 2007).

We intend to leverage Industry 4.0 technologies in these processes. Industry 4.0 is defined as a new industrial phase in which there is an integration between production operational systems and information and communication technologies (ICT) – particularly the Internet of Things (IoT) – forming cyber-physical systems (CPS). This new industrial phase is affecting the rules of competition, industry structure, and customer demands. It is also changing the rules of competition as companies' business

DOI: 10.1201/9781003740148-7

models are being reshaped by adopting the concepts of the Internet of Things and the digitalization of factories.

Now the questions that arise are:

What impact have Industry 4.0 technologies had on manufacturing industries?
How can Industry 4.0 be used to improve different sectors of the industry?

Today, many retailers have realized the need to integrate their existing channels to enrich the customer value proposition and improve operational efficiency. As a result, there is a new focus on omnichannel retailing "with the aim of providing consumers with a seamless shopping experience across all available shopping channels". So, given the changing needs of customers, there is a need for manufacturers to focus on the efficiency of multi-channel sales.

Also, the unique role of customers and their participation in each stage of new product development helps companies to offer new products that better meet customer needs and, ultimately, improve the performance of companies' new product development. So, manufacturers must definitely think about improving customer interactions.

Also, digital technologies allow companies to offer new digital solutions, such as internet-based services embedded in products, to customers. From an operational perspective, digital technologies such as CPS offer the potential to reduce setup times, labor and material costs, and processing times, and thus higher productivity of production processes. So, considering the above, the need to utilize the technologies of Industry 4.0 is quite evident, because it allows manufacturing industries to do their work more efficiently and at a lower cost, and to have better management over the various production and management sectors. And this issue is absolutely necessary in the competitive market of this era, especially in our country, which is under heavy sanctions.

Main Objective:

Using Industry 4.0 technologies in new product development and multi-channel sales in the B2C model.

Specific Objectives:

Understanding Industry 4.0 and related technologies.
Understanding product sales and marketing through multiple channels.
Determining how to apply Industry 4.0 technologies in new product development.
Determining how to apply Industry 4.0 technologies in multi-channel sales.
Determining how to apply Industry 4.0 technologies to improve customer interaction

Practical Objective:

Create a model for applying Industry 4.0 technologies in new product development and multi-channel sales in the B2C model for use in manufacturing industries.

Research Questions:

How to use Industry 4.0 technologies to improve new product development?
How to use Industry 4.0 technologies to improve a multi-channel sales strategy?
How to use Industry 4.0 technologies to improve customer interaction?
How to use Industry 4.0 technologies to increase customer satisfaction?

7.2 BACKGROUND

Researchers conducted a study to identify various factors using the meta-synthesis method and designed a process model based on the interpretive structural modeling method. The findings from the meta-synthesis of the research showed that 29 factors affect new product development. The process model also clarified that different factors are located at 12 levels, among which environmental, cultural, and managerial factors are very important as key and underlying factors.

In another study, whether customer participation creates a synergistic or destructive effect on the market performance of the new product at each stage of new product development was investigated. The results showed that customer involvement in the ideation and development stages and in both the development and launch stages has synergistic benefits, while customer involvement in both the ideation and launch stages does not create any additional gains (Fazlollahtabar and Aghasi, 2012).

An article provides a structured overview of some of the most significant changes in the new product development process of companies (new product development). Starting with the well-known new product funnel, the article explains how the multi-channel environment and the technologies that enable it affect the speed and execution of each stage of new product development. It illustrates the changes described with examples, including consumer packaged goods, consumer technology, and fashion.

In a study, the modeling of a four-level integrated multi-channel supply chain for a seasonal product under random demand was investigated. They modeled the data in decentralized and centralized decision-making modes. After determining the optimal values of the decision-making variables in each mode, they tested the results using real data and evaluated them numerically. Finally, they concluded that the average profit of the members and the total profit of the supply chain increase with an increase in the final selling price and the production capacity of the product, but the growth rate of profit with an increase in the selling price is higher than the growth rate of profit with an increase in maximum production.

A paper presents a forecasting method for an omnichannel retail supply chain database using clustering algorithms and artificial neural networks to reduce demand-related uncertainty. Through the application of this method in a Brazilian retailer, its performance is analyzed, and it is concluded that for multi-channel retailers, demand forecasting needs to combine pattern analysis and forecasting methods to improve forecast accuracy and thus be able to reduce uncertainties in the multi-channel retail supply chain.

Researchers conducted a four-stage Delphi study with 18 retail experts to identify key trends, major challenges, important technologies, and key customer touchpoints that will emerge in multi-channel retail over the next ten years. Using both qualitative and quantitative data analysis techniques, it first extracted open-ended predictions from experts, then transformed and aggregated these predictions into final statements, finally obtained expert ratings of these statements, and analyzed the changes in experts' ratings between two successive stages. Based on this approach, it obtained four general themes of key insights: Future competition in the retail industry will be based on comprehensive customer experiences; Omnichannel retailing requires the development of human capabilities and a change in organizational mindset; Physical stores will become key destinations for unique sensory shopping experiences; and Omnichannel retailing improves operational efficiency (Mahdavi et al., 2007).

Another study examined the influencing factors of customer channel choice intention in the multi-channel retail environment. They collected data through a survey of customers (most of whom are online shoppers) from the United States and the United Kingdom, followed by preliminary analysis through confirmatory factor analysis to assess the characteristics of the measure, and then, after confirming a suitable measurement model, they tested the interrelationships between the variables latent in the model through SEM.

They found that channel transparency, channel convenience, and channel uniformity had a positive effect on customer perceived behavioral control. Furthermore, they found that channel transparency and uniformity helped reduce customer perceived risk, while channel convenience had no significant effect. Higher product price increased the effect of channel transparency, convenience, and uniformity on reducing customer perceived risk. Furthermore, they found that customer behavioral control and channel price advantage had a positive effect, and perceived risk had a negative effect on customer perceived risk (Mahdavi et al., 2009).

Researchers in the Industry 4.0 conducted a study with the aim of improving the food industry using Industry 4.0 technologies. To achieve this, they first extracted 20 cases of Generation 4.0 industry technologies through literature extraction, and in the second stage, they identified 13 cases of Generation 4.0 industry technologies for explanation in the food industry using the Fuzzy Delphi approach. Then, they used the Revised DEMTEL to investigate the interactive relationships of Fourth Generation Industry Technologies. They were finally able to identify a number of Generation 4.0 industry technologies as influential factors in the food industry.

Using secondary data from a large survey of 27 industrial sectors representing 2,225 Brazilian manufacturing companies, they observed how the use of different Industry 4.0 technologies is related to expected benefits for product, operations, and side effects. They then used regression analysis to show that some Industry 4.0 technologies are promising for industrial performance while some emerging technologies are not, which contradicts conventional wisdom (Fazlollahtabar et al., 2010).

In a study, researchers aimed to identify different adoption patterns of two layers of Industry 4.0 technologies: basic technologies and advanced technologies. To do this, they surveyed 92 manufacturing companies to study how these technologies are implemented. Then, they combined two methods: hierarchical cluster analysis (HCA) followed by demographic analysis and independence tests. Their results support the hypothesis that smart manufacturing plays an important role in Industry 4.0 and is closely related to smart products. They also show how other advanced technologies complement smart manufacturing. Also, according to their findings, companies with advanced levels of Industry 4.0 implementation tend to adopt the most advanced technologies and not have a specific subset. Another paper identified the fundamental changes needed in the new product development process and proposes a new structure to capture the value from adopting Industry 4.0 technologies. This study was conducted in the field of the tooling industry, and the proposed changes were approved by industry experts (Fazlollahtabar, 2024).

In the literature, researchers presented a digital supply chain that includes traditional supply chain actors and structures as well as new and emerging elements inherent in Industry 4.0. This model was based on and supported by the results of a two-part literature review. The first part of this review identified the most important conceptual models in the field of supply chain management published from 1989 to 2019 and revealed the lack of a comprehensive digital supply chain model. The second part of this review systematically examined the status of Industry 4.0 and identified a summary of the emerging elements and technology structures used in the nascent digital supply chain.

In this section, first, in the theoretical foundations of the research, we found a complete view of the concepts of new product development, multi-channel sales, and Industry 4.0 by reviewing various theories. In the second part, the research background, we reviewed domestic and foreign research on the concepts of new product development, multi-channel strategy, and Industry 4.0. Finally, it can be said that good research and studies have been conducted on each of the main concepts of this thesis in a specialized manner, but unfortunately, not much attention has been paid to these topics within the country. Given that these topics are very important and influential in the industrial future of the country, there is a need to pay more attention to this sector and conduct more scientific research. However, if we look at it from an international perspective, a lot of research has been conducted on each of these concepts separately, but not enough scientific research has been conducted on the interaction of these issues with each other, such as the role of Industry 4.0 in the development and improvement of multi-channel sales and marketing, as well as the interaction of all three general concepts of the research topic, namely new product development, multi-channel sales, and Industry 4.0.

7.3 METHODOLOGY

Now that we are fully familiar with the concepts presented and have reviewed the background of various research related to the subject, we will move on to research modeling. To do this, we will first introduce the tools that we will use for modeling, and then we will draw and present the research model and explain it in full. After that, we will define the indicators and variables, and finally we will introduce the research data collection tool.

As mentioned, new product development includes several stages of decisions and processes during which we develop an idea and finally turn it into the desired product. Also, in the competitive business environment these days, customer participation in the development of a company's new product is more important than ever, and companies need to participate in the various stages of new product development in order to progress and develop.

We also talked about multi-channel strategy and saw that in the current era of sales and marketing, it has undergone many changes. In the past, the customer experience began and ended in a store, but today it has become a journey across multiple channels, and the current challenge for businesses is how to create an effective and consistent customer experience across all of these channels. Companies that use a strong and appropriate multi-channel strategy will have many opportunities.

Now it is time to examine what capabilities Industry 4.0 will bring to these two sectors, as well as other areas such as supply chain and customer interaction, etc. So, to do this, we want to design a model for producing a new product and selling it through multiple channels using Industry 4.0.

Indicators:

Independent variables: Quality – System efficiency
Dependent variables: Customer satisfaction – Customer demand – Profit
Send feedback
Side panels
History
Saved
Revenue: It is the opportunity cost of consumption and savings by a business entity over a specified period of time, generally expressed in monetary terms.
Customer satisfaction: The number of customers or percentage of total customers whose reported experience with a company, its products, or services (ratings) exceeds established satisfaction targets. It is a measure of how well the products and services provided by a company meet or exceed customer expectations.
System efficiency: It represents the level of performance that uses the least amount of input to achieve the greatest amount of output. This often specifically involves the ability of a particular program of effort to produce a specific result with the least amount or amount of waste, cost, or unnecessary effort.
Quantity of demand: The amount of a good that consumers are willing and able to purchase at various prices during a given period.
Quality: Suitability for the intended purpose (fitness for purpose) while meeting customer expectations.
Profit: The difference between a business unit's income from its outputs and the costs of its inputs.

7.3.1 Data Collection

In this research, library data collection method has been used as a conventional and reliable method to clarify and explain the foundations and topics of the research background and to obtain the information required to design the proposed model. These sources can be categorized into the following groups:

- Valid Persian scientific articles
- English articles in the databases: sciencedirect, emerald, etc.
- Some specialized online books in the fields of information technology and management
- Review and study of some Persian theses available on the web

7.3.2 Research Model

In this section, using the information we obtained from various sources, we will first briefly examine the various applications of Industry 4.0 in manufacturing industries and then, using them, we will design a model for applying Industry 4.0 in manufacturing industries.

Application of Industry 4.0 in different parts of the system:

Supplier: Automatic ordering of requirements and synchronization of supplier and manufacturer. By integrating data throughout the supply chain, in real time and without human intervention, delivery waiting times can be significantly reduced, and load and inventory management can be optimized to enhance the agility of the entire chain.

Using data mining and machine learning techniques, the performance management system can identify the root causes of performance deviations by analyzing big data and automating planning systems.

Production and development:

- Optimization: Using smart devices that can work more accurately and with less energy consumption
- Reducing production time and eliminating time constraints by eliminating human intervention in some production lines
- Fast and automatic equipment troubleshooting, timely notification, and prevention of collateral damage
- Modern system for using data in real time for preventive operations and remote maintenance
- Data mining of data sent by sales channels, analyzing them, and using useful data in improving the current product and developing new products

Sales channels:

- Creating coordination between sales channels: inventory, etc.
- Smart warehousing and trucks on their way to the warehouse send their location and time of arrival to the smart warehouse management system, and the system selects and prepares their dock for optimization and timely and correct delivery.
- Troubleshooting and accurate checking of goods before shipping and when shipping to the customer.
- Understanding customer needs and feedback.

By using technologies embedded in products, data feedback is received for troubleshooting and also for developing new products, and is sent to the production and development departments.

Customers:

- Detailed training on how to use the product using augmented reality tools
- Real-time positioning of the shipments by customers
- Accurate calculation of the delivery time of the product according to the weather conditions, traffic conditions, etc.

In Figure 7.1, the designed system is depicted using an entity relationship (ER) diagram.

In Figure 7.1, which is the ER diagram of the system we designed, we have described the different entities and their attributes so that you can become familiar with the features of each part of the system.

Supplier: The supplier of raw materials for production includes the attributes ID (primary key), name, address, phone number, and email.

Material: Raw materials for production that the manufacturer buys from the supplier including the attributes: code (primary key), name, type, description, price, weight, size, and date. And also the attribute supplierID, which is a foreign key and is used to connect the raw materials with the supplier.

The relationship between the supplier and the raw materials is of the many-to-many type. That is, each supplier supplies several raw materials, and each raw material can be supplied by several suppliers.

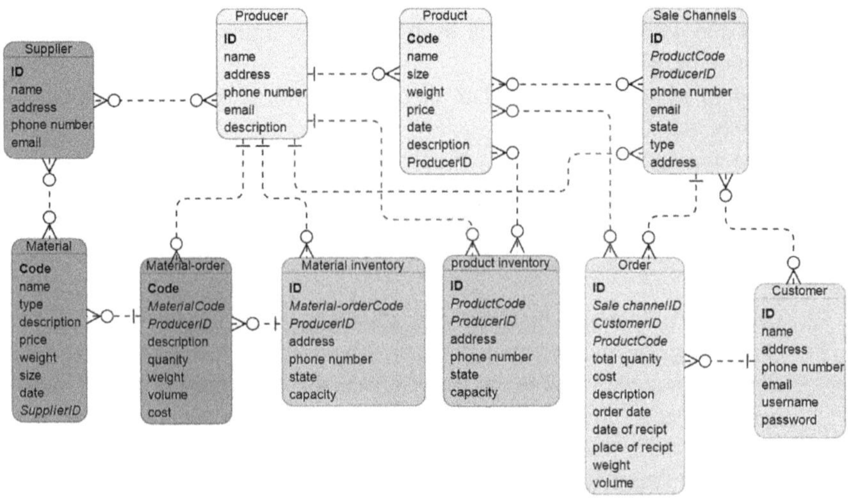

FIGURE 7.1 Entity relationship diagram.

Producer: After purchasing raw materials from the supplier, it starts producing products, which include the attributes: ID (primary key), name, address, phone number, email, and description.

The manufacturer has a many-to-many relationship with the supplier and a one-to-many relationship with the sales channels, raw material warehouse, raw material order, and product warehouse.

Material order: The manufacturer orders the required raw materials from the supplier, which includes code (primary key), description, quantity (countable materials), volume and weight (non-countable materials), and cost. And also the attributes MateriaCode and ProducerID are foreign keys.

This entity has a many-to-one relationship with the manufacturer and raw material warehouse and a one-to-many relationship with the raw materials.

Material inventory: The manufacturer stores the purchased raw materials there, which includes the attributes: ID (primary key), address, phone number, status, and capacity. And also Material-orderCode and ProducerID are foreign keys.

This entity has a many-to-one relationship with the manufacturer and a one-to-many relationship with the raw material order.

Product: A product produced by a manufacturer and includes the attributes code (primary key), name, size, weight, price, date, and description. ProducerID is also a foreign key.

The product has a many-to-one relationship with the manufacturer and a many-to-many relationship with the sales channels, product warehouse, and order.

Product inventory: The location where products are stored after production, which includes the attributes ID (primary key), address, phone number, status, and capacity. ProductCode and ProducerID are also foreign keys.

This entity has a many-to-one relationship with the manufacturer and a many-to-many relationship with the product.

Sale channels: The channels through which the manufactured products are sold, which include the attributes: ID (primary key), phone number, email, status, type (offline, online), and address. ProductCode and ProducerID are also foreign keys.

This entity has a many-to-one relationship with the manufacturer, a one-to-many relationship with the order, and a many-to-many relationship with the product and the customer.

Customer: A buyer who buys the products produced through sales channels, which includes the attributes: ID (primary key), name, address, phone number, email, username, and password (for online purchases).

This entity has a many-to-many relationship with the sales channels, and a one-to-many relationship with the order.

Order: An order to purchase goods by a customer in one of the sales channels, which includes the attributes of total quantity, cost, description, order date, date received, and place of receipt (online purchases), weight, and volume (postal deliveries). Also, Sale channelID, CustomerID, and ProductCode are foreign keys.

This entity has a many-to-one relationship with the customer and sales channels, and a many-to-many relationship with the product.

Use Case of the Proposed Model:

Use Case shows the interactions between the various elements of the system. Four actors are considered, namely: supplier, production and development, sales channels, and customer.

On the left side of the image is the supplier. In our designed system, which uses Industry 4.0 technologies, the supplier exchanges information with the manufacturer intermittently, with most of the information flowing from the manufacturer to the supplier. After detailed analysis, this information is divided into several categories, including: the resources required by the manufacturer and the products that the manufacturer intends to produce, as well as complete information about the production systems that the manufacturer uses to produce the products. This information is used to evaluate and coordinate the supplier and the manufacturer correctly and accurately, in order to optimize and accelerate the supply chain and eliminate inadvertent errors that cause losses to the manufacturer.

In the next section, we will first look at the relationship between sales channels and customers. In our sales system, we have used the omnichannel sales capability, and by utilizing Industry 4.0, there is precise and complete coordination between the different sales and marketing channels.

After marketing, the customer makes a purchase through one of the online stores, physical stores, or social networks. After entering the store (online or physical), the customer can receive complete information about the store's products through augmented reality. After choosing the product of his choice, a file is created for him with the customer's consent, so that comprehensive and accurate monitoring of the customer and how to use the products can be formed in the future, and new and complete information about new products is also provided to him. After completing the forms, the customer selects the location to receive the product. If the customer is buying from an online store or social networks, they can choose the option to receive it on site or in one of the branches of physical stores. And if he is buying in a physical store, he can receive the product right away or request that the product be delivered to the customer's designated location.

After that, the customer must specify the payment method. In the online store and social networks, there is the ability to pay by card or cash on delivery. And also in the physical store, the customer can pay online or with bank cards.

After payment, the product is delivered to the customer, and the customer can learn through augmented reality a detailed and complete training on the correct way to use the products. Also, with the customer's satisfaction, from the moment of purchasing the product, using technologies embedded in the product, the way the product works, and how the product is used by the customer is monitored moment by moment, and real-time information is sent to the sales channel systems. Also, communication is established with the customer during specified periods, and their comments and feedback are recorded in the system.

This stored information, which includes sales information, customer information, and warehouse information, is sent immediately to the production unit system.

In the next section, we will look at the interactions between the production system and sales channels. The initial part of the tasks of the production department system is the intelligent management of warehouses using Industry 4.0, which is responsible for the delivery and clearance of raw materials as well as manufactured products in relation to suppliers and sales channels.

Also, the production department systems monitor all sales and marketing channels accurately and in real time, and the data recorded in the systems of these channels is received in real time. After careful analysis of this data, it is divided into five sections, which include customer information, sales information, production systems information, warehouse information, manufactured product information.

In the next stage, this data is used for troubleshooting and rapid improvement in sales and marketing channels, as well as in the warehouse sector, and especially in the production and product development sector.

Activity Diagram of the Proposed Model:

Our information flow cycle starts from the production department and ends in the production department. This cycle repeats periodically, and in each iteration, the information obtained after analysis is used to improve the system in different departments and also to troubleshoot the system in different departments.

Our information analysis is done using data mining and factor analysis. Factor analysis in this system is used to simplify the data transferred between different systems and also to identify suitable variables for subsequent analysis.

Systems also use data mining in various cases. For example, identifying the best product and the best sales channel, predicting customer behavior, determining the best advertising method based on customer knowledge, predicting product sales from different channels, etc.

Systems also automatically select the best algorithm for the data available to them.

In the following, we will learn more about data mining and factor analysis.

7.3.3 Modeling

In this section, we will learn more about the methods that the system uses to analyze information obtained from different systems.

Data Mining Method: Data mining is a non-trivial process for identifying valid, novel, potentially useful, and ultimately understandable patterns in data. Data mining is the search for valuable information in large volumes of data. It is the process of extracting non-trivial implicit, unknown, and potentially useful information, such as knowledge rules, constraints, and order, from data stored in repositories using pattern recognition technologies as well as statistical and mathematical techniques.

Depending on the analysis goals, different types of data mining operations can be used. In general, these data mining operations are used to describe the general characteristics of the database or to make inferences on the data for the purpose of prediction.

In general, there are three main steps in data mining: data preparation, data reduction, and finally, the search for valuable information. However, specific approaches vary from company to company and researcher to researcher. For example, past works proposed the following steps:

1. Retrieving information from a large database.
2. Selecting the relevant subset for the task.
3. Deciding on the appropriate sampling system, cleaning the data, and dealing with missing fields and records.
4. Applying appropriate transformations, dimensionality reduction, and predictions.
5. Fitting the models to the pre-processed data.

7.3.4 Factor Analysis

Factor analysis operates on the concept that measurable and observable variables can be reduced to a smaller number of latent variables that share common variance and are not observable, which is known as dimensionality reduction. Large data sets consisting of multiple variables can be reduced by looking at "groups" of variables (i.e., factors) – that is, factor analysis collects common variables into descriptive categories. Factor analysis is useful for studies that involve a few or hundreds of variables, questionnaires, or a set of experiments that can be reduced to a smaller set, to understand basic concepts and facilitate interpretations. It is easier to focus on a few key factors than to treat many variables as insignificant, and therefore factor analysis is useful for placing variables into meaningful categories. Many other applications of factor analysis include data transformation, hypothesis testing, mapping, and scaling. Now that we have become familiar with data mining and factor analysis, we will analyze in the next chapter how to analyze various data collected by the system using these tools and methods.

7.4 IMPLEMENTATION STUDY

In the previous section, a model was designed, and the mechanism of the different parts was described, and it was also shown what data is exchanged between different systems and what actions the systems take with this data. Now, in this section, after generating different random data, we will run the model that we designed in the previous chapter numerically and then interpret the results obtained to show the mechanism of the system and the designed model. And our goal in this chapter is not just the result but also to get acquainted with some of the calculation and analysis methods that the systems in the model designed in the previous chapter perform.

Now we want to test the proposed model that we designed in the previous section. For this purpose, first we randomly generate data related to each of the different systems, and then we extract the required information using data mining and factor analysis methods.

Our production data is as follows:
Customer data:
This data includes the following:

1. Customer ID: The number of customers in this data is 500, identified by the numbers 1 to 500.
2. Purchased product number: The product that each customer has purchased. In this dataset, we have 3 different products, identified by the numbers 1 to 3.
3. Customer purchase channel: The channel through which the customer has purchased the product.
   We have 3 different sales channels: 1: Online store, 2: Physical store, 3: Social networks
4. Customer satisfaction with the product: The customer's satisfaction with the product is requested from the customer through a questionnaire, and the scores will be from 1 (very dissatisfied) to 5 (completely satisfied).
5. Customer satisfaction level with the sales channel: The customer's satisfaction level with the sales channel is requested from the customer through a questionnaire, and the scores will range from 1 (very dissatisfied) to 5 (completely satisfied).
6. Customer satisfaction level with timely delivery of goods: The customer's satisfaction level with timely delivery of goods is requested from the customer through a questionnaire, and the scores will range from 1 (very dissatisfied) to 5 (completely satisfied).

Sales Channel Data:
This data includes the following:

1. Channel Type: We have 3 different sales channels: 1: Online Store, 2: Physical Store, 3: Social Media.
2. Number of Personnel: Number of personnel present in each of the sales channels.
3. Amount of Capital Spent: Amount of capital spent for each of the sales channels.

Product Sales Data:
This data includes the following:

1. Product Number
2. Price: Price of each product
3. Transportation Cost: Amount of capital spent on transporting the product (from the company to the warehouse, from the warehouse to the consumer).
4. Marketing Cost: Amount of capital spent on marketing each product.

Product Warehouse Data:
This data includes the following:

1. Product Number
2. Occupied Space: Percentage of warehouse space occupied by each product
3. Weight: Weight of each unit of the product
4. Storage Cost: Amount of cost spent on storing and maintaining the product in the warehouse (insurance, taxes, etc.).

Production system data:
This data includes the following:

1. Device number: The number of production devices, of which there are 13 in the data set, and are indicated by the numbers 1 to 13.
2. Maintenance cost: The amount of money spent on the production devices (fuel, repairs, etc.).
3. Number of errors: The number of system errors during 1 month.
4. Number of products produced: The number of products produced by the device during 1 month.
5. Number of supervisors: The number of supervisors for each production device.
6. Product number: The number of products produced by each production device.
7. Production time: The time required to produce output.

Raw material warehouse data:
This data includes the following:

1. Raw material number: Raw material number, which is 17 in total and is identified by the numbers 1 to 17.
2. Product number: Product number associated with each raw material.
3. Monthly purchase cost: Monthly purchase cost of each raw material.
4. Occupied space: Percentage of warehouse space occupied by each raw material.
5. Weight: Weight of each unit of raw material.
6. Storage cost: Amount of cost spent on storing and maintaining raw materials in the warehouse (insurance, taxes, etc.).

7.4.1 Data Analysis

Now that we have prepared the various data, we want to analyze this data and check which is the best product produced by the company and which is the worst product. And also check which sales channel performed better and which channel performed worst. For this purpose, first we analyze the data related to customer satisfaction through a graph and check the results, and then we perform the analysis by taking into account all the columns of the data using factor analysis, and finally we analyze and check the results obtained.

Customer Information Analysis:
Here, we want to combine the satisfaction scores that customers gave to different departments in different groups and discover the possible relationships between them. For this analysis, SPSS Modeler software was used.

1. Combining product number and product satisfaction columns
   From the results, product 2 has the best satisfaction, product 1 has the worst satisfaction, and product 3 is in the middle. It can also be seen that product 2 has more sales than the others and is almost more popular.
2. Combination of purchase channel columns and satisfaction level with purchase channel
   From the results, channel 3 (social media store) has the best satisfaction, but it is the least popular channel, and channel 1 (online store) has the lowest satisfaction level, but it is the most popular channel, so it is necessary to plan better for it as soon as possible.
3. Combination of columns of satisfaction with the purchasing channel and satisfaction with timely delivery

From the results, the higher the satisfaction with the timely delivery of goods, the greater the satisfaction with the sales channel, so satisfaction with timely delivery of goods has a direct effect on satisfaction with the purchasing channel.

The remaining combined columns have neither a logical nor a relative relationship. So we end this method here and start the next part of the analysis.

7.4.1.1 Applying Factor Analysis

Now we want to perform the analysis using factor analysis. This method is implemented by coding in Python. In this method, we involve all meaningful columns in the problem analysis process. First, we combine all data tables together. Then we removed meaningless columns in the data analysis process such as customer number, etc., which ultimately left 20 columns. After that, we ran the factor analysis code, and the system determined that the number of factors could be reduced to 7 based on the data. As shown in Figure 7.2, given that we have 7 eigenvalues greater than 1. So the number of factors can be reduced to 7.

From the results, all eigenvalues are displayed; the first 7 factors in the eigenvalues are higher than 1. Also, for each of the 7 selected factors, the values of the sum of squared times are significant, which are used in the calculations. Now, to analyze

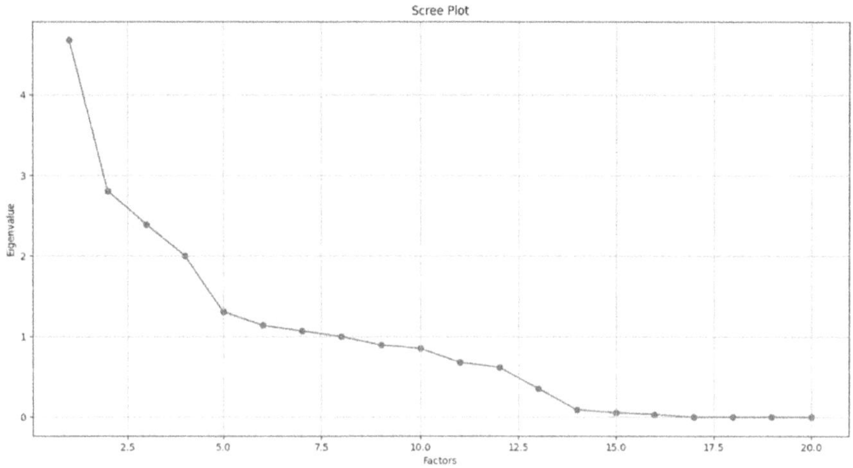

FIGURE 7.2 Factor analysis plot.

the data, first calculate the average of all 3 columns related to customer satisfaction, and then calculate the percentage of the best being superior to the worst for all seven existing factors. And then these numbers were added to the corresponding channels and products, and thus the best and worst were calculated. The final output using factor analysis is given below:

Best product= 2, Best-selling channel= 3, Worst product= 1,
Worst-selling channel= 2.

From the results, the final output can be seen that product 2 and channel 3 (social networks) were calculated as the best, and product 1 and channel 2 (online store) were calculated as the worst.

7.4.2 Analysis of the Results

As can be seen from the above analysis method, both methods published the same result.

Of course, this was only a corner of the calculations that the systems perform in the designed model and based on different data and different requests, they perform different methods to analyze and examine the data exchanged in the system. Also, over time, the size of the data set continues to grow larger and larger, and there will be a need for factor analysis to make data analysis easier and faster by reducing factors, which will allow urgent problems and requests to be addressed in real time and reduce the pressure on the systems.

In this section, the analysis of the data exchanged in the model designed in the third section was discussed to show the mechanism of the system and an example of the system's work in dealing with the collected data. At first, the data that was generated randomly was displayed and introduced in different tables. Then, with the aim of discovering the products and channels that performed better and worse in this data set, we analyzed this data in two different ways. And finally, we encountered similar results from both analyses.

7.5 CONCLUSIONS AND SUGGESTIONS

As seen in the previous sections, in the first section we overviewed the problem and mentioned the objectives and questions of the research. Then, in the second section, which was called the background, the theoretical foundations of the research were first stated, and then we had a review of the research background. After that, in the third section, using the knowledge we gained in the research process, we implemented a model and explained its general performance, and then, in the fourth section, we tested this model using random data and explained two methods of analyzing this model.

In the first section, some questions were stated that were a path for conducting the research. Now, in the final section, after answering those questions, we will examine the results obtained.

1. How to use the technologies of Industry 4.0 to improve the development of new products?
   According to the knowledge we gained in the research process and according to the model we designed, we found that production systems, by collecting and analyzing data related to customer satisfaction with products, sales volume, and other matters, can play a significant role in identifying customer tastes and changing and upgrading products based on customer tastes, as well as identifying products that had low sales and troubleshooting and changing them to improve their sales.
2. How to use the technologies of Industry 4.0 to improve the multi-channel sales strategy?
   According to the knowledge we gained in the research process and according to the model we designed, we found that systems, by collecting and analyzing data related to sales channels, including: customer satisfaction with sales channels, customer satisfaction with timely delivery of goods, costs related to channels, etc., can identify the strengths and weaknesses of each sales channel and then choose an appropriate strategy to improve or strengthen sales channels.
3. How to use 4.0 industry technologies to improve customer interaction and increase customer satisfaction?
   According to the knowledge we gained in the research process and according to the model we designed, we found that systems using technologies such as the Internet of Things and augmented reality can form much closer relationships between customers and manufacturers, which will also increase the manufacturer's understanding of customer needs, as well as real-time information about product-related problems and immediate troubleshooting by manufacturers, as well as complete knowledge of the products purchased by the customer, as well as real-time information about changes in current or

new products. Ultimately, these will increase customer trust and, of course, increase customer satisfaction with the manufacturer.

As a result, considering all the answers given to the questions, we realize the extraordinary and fundamental role that Industry 4.0 technologies can play in various manufacturing industries, and it becomes clear that various industries in the world, and especially in Iran, need to take action to implement these technologies as soon as possible.

Given that Industry 4.0 has not yet found its way in Iran and only a handful of sectors, and not completely, use these technologies, the ability to implement the designed model in a real system was not there, and we were forced to use random data in the process of analyzing information.

Some of the suggestions follow here:

1. Given the rapid growth of technology and corrective changes with respect to these technologies, adapting as quickly as possible and equipping manufacturing industries with 4.0 generation industry technologies as quickly as possible can be a fundamental and important step to win in this competitive market and also increase production and reduce losses, and of course reduce production costs, which will ultimately increase profits and productivity for every company and manufacturing plant.
2. If we look at this issue from a national perspective, we see that our country, unfortunately, has not yet benefited much from the technologies of Industry 4.0 in various industrial sectors, which could soon lead to the victory of competitors in various industries and the seizure of markets that were previously in our country's hands. And of course, given the recent sanctions and the need to realize a resistance economy, it is necessary to use these technologies in industries as soon as possible in the country so that we can seize new economic markets in the world.
3. Given the increasing necessity of customer participation over time, manufacturers must constantly seek to create new solutions for more effective interaction with customers.
4. Given the change in customer purchasing trends, it is necessary for sellers to move away from their traditional perspective and move toward multichannel sales.

Suggestions for future research:

1. A detailed examination of the infrastructure required for the correct and optimal implementation of Industry 4.0 in the country.
2. Investigating how to build trust between customers and 4.0 generation industry technologies, as well as building trust between manufacturers and 4.0 generation industry technologies in the country.
3. Investigating how to prepare the country culturally for accepting 4.0 generation industry technologies.

8 Cloud Processing in Sustainable ERPs for Small and Medium Enterprises

8.1 INTRODUCTION

Small and medium-sized organizations have more limited resources and budgets than large organizations and are more sensitive to costs. Studies show that cloud computing plays a supporting role in improving the costs of small and medium-sized organizations. Given that user needs are increasing and are often not met by individual services, a composite service should be achieved by combining a number of basic services to suit the user's desired features. Since our problem is to find an optimal combination, we will use the dimensions of sustainability and its indicators to find an optimal architecture. A traditional enterprise resource planning system has a philosophy of only increasing the organization's profitability, while the philosophy of our Sustainable Enterprise Resource Planning system includes components of greater profitability along with better environmental protection and social benefits. Therefore, choosing a basic service that suits the user's desired features is an important issue. The goal of this research is to combine these services based on service design principles in such a way as to provide a cloud-based ERP2 information system whose optimization is based on sustainability indicators, including three dimensions – environmental, social, and economic – for small and medium-sized organizations.

Processes have value from a business perspective and pursue a specific goal to ultimately produce a product/result. The role of technology here is to automate and optimize the execution of these processes. However, on the other hand, these information systems have heterogeneous technologies and standards. One of the important problems for process automation and management is how to implement a process that is related to heterogeneous information systems in an integrated and comprehensive manner.

In recent years, service-oriented architecture has succeeded in solving several problems with its high flexibility and integration capabilities. It should be noted that one of the principles of service-oriented architecture is the independence of organizational services from technologies and platforms (independent of platform), so that the infrastructure of this architecture is based on the interaction and cooperation of a set of diverse information systems, each of which has its own technology and standards,

DOI: 10.1201/9781003740148-8

but where there is a need for interaction between systems in the form of providing/receiving services, these services are based on a global and accepted standard that allows other systems to call these services. Service-oriented architecture allows the organization's information systems to have their own technology and implementation, provided they follow an accepted standard for providing and receiving services with other systems. The concept of "service" also expresses this issue. A service in the cloud is a mechanism that can provide two or more features that allow a user to comply with the restrictions and rules defined by the provider. In complex business processes where a single service cannot perform all the required tasks, to overcome this complexity, different services must be integrated with each other.

Small and medium-sized enterprises require special attention when it comes to business strategies toward sustainable development, as the business case is not the same as that of large companies. Furthermore, tools designed to support sustainability in SMEs should take into account that these companies have different resources and assets than large companies. However, small and medium-sized enterprises (SMEs) systematically lack capacity, productivity, a trained workforce, and managerial and entrepreneurial skills. Furthermore, half of SMEs lack access to finance, with a low-cost gap of $2.1 to $2.6 trillion. Therefore, this research has paid attention to these aspects in the selection of services and design for this type of company.

In order to address the above challenge, this research is divided into seven main parts. In the first part, definitions and characteristics of service-oriented architecture are presented, and the composite service architecture is examined. In the second part, we are introduced to enterprise resource planning systems based on cloud computing. In the third part, the dimensions of sustainability and various indicators for measuring it in the organizational system are analyzed. In the fourth part, small and medium-sized organizations are described, and cloud computing and its appropriate architectures are discussed. The strengths of an optimal information system for such organizations are mentioned. In the fifth part, a sample scenario and its architectural model are presented and discussed. In the sixth part, the proposed model and architecture are presented and evaluated. In the last part, conclusions are drawn.

We have come to the necessity of sustainability as a key differentiating factor in today's business world, and our goal is to achieve a Sustainable Enterprise Resource Planning system for small and medium-sized companies because these companies face financial constraints to achieve a traditional enterprise resource planning system.

The goal of this research is to identify the processes of small businesses and find an approach to increase the competitive advantage of these companies by designing composite web services using our cloud environment, which ultimately provides us with an optimal management system, given that one of the problems of these businesses is resource scarcity. The enterprise resource management system in the cloud infrastructure provides a management system that integrates and consolidates various information processing facilities. Therefore, our emphasis in this research is to provide a solution for the cooperation of small and medium-sized companies in the supply chain.

In fact, some of the general objectives of this research include:

- Providing a solution to solve the problems of small and medium-sized businesses

- Reducing the cost of processes in these businesses
- Integrating the goals of these businesses to meet customer needs and compete with competitors
- Familiarizing these companies with sustainability philosophies and their competitive advantage
- Providing a model in terms of sustainability indicators so that the organization can remain successful despite internal and external pressures
- Implementing an enterprise resource planning system based on cloud computing
- Reducing the cost of implementing an ERP system for small and medium-sized organizations
- Improving business processes with regard to the optimal combination of services
- Improving the quality of composite service by selecting appropriate services

In this study, we seek to answer the following questions:

- How is it possible to link sustainable ERP systems with the SME sector?
- What is the role of small and medium-sized organizations in sustainable development?
- What is the impact of the enterprise resource planning system on small and medium-sized companies and organizations?
- What are the optimality indicators for the composite service architecture?

In this study, the following assumptions are considered for modeling and implementation:

- ERP is the integrated management of the organization's main processes in real time with the help of technology and software.
- S-ERP is derived from the ERP concept and is a type of Sustainable Enterprise Resource Planning system.
- Economic, social, and environmental dimensions are considered in obtaining the proposed model.

This model is suitable for small and medium-sized companies and distributed companies.

8.2 ENTERPRISE RESOURCE PLANNING SYSTEM

ERP systems are comprehensive software packages that provide solutions that help integrate all processes and functional areas in a business. The main and primary advantage of these systems is the creation of a common integrated software platform for integrated processes in the relevant business. These systems have the ability to provide services to companies of any size and volume of activity and any geographical spread. Numerous definitions have been provided for ERP systems, and a few of these definitions are mentioned below.

- ERP is a comprehensive system that attempts to integrate all the functions and departments in an organization using a single computer system so that it can meet the specific needs of these departments.
- ERP is a thinking technology and system for more effective management of various resources in an organization, which is done by automating and integrating all processes, thereby increasing the efficiency of the organization and increasing customer satisfaction.
- ERP is an information technology-based solution that puts the organization's resources under the control of managers at different levels of the organization through an interconnected system with speed, accuracy, and high quality to appropriately manage the organization's planning and operations process.

8.2.1 Sustainable Resource Planning System (S-ERP)

Sustainable development is an organizing element that ensures the sustainability of non-renewable resources. These are limited resources that are essential for the life of future generations on the planet. Sustainable development is a process that envisions a desirable future for human societies in which living conditions and resource use meet human needs without compromising the integrity, beauty, and stability of vital systems. Sustainable development is an important new perspective in public management and policymaking that has emerged widely outside the United States. The concept attempts to consider more clearly the future consequences of current behaviors. Sustainable development addresses a variety of issues, including the impact of greenhouse gases, climate change, ozone layer depletion, land degradation, depletion of non-renewable resources, and urban air pollution.

The concept of sustainability usually encompasses various domains such as ecology, economics, politics, and culture, and achieving it requires considering the inherent or desirable characteristics in all of these domains. A company that is committed to a sustainable future benefits greatly from the perspective of social image, stakeholder capital, and operational efficiency. Implementing production and distribution systems that reduce energy consumption and pollutant generation, prevent waste generation or reuse through recycling, and reduce or minimize overuse of primary resources (natural resources) is not only the right ethical act today but also a competitive advantage.

8.2.2 Global Value Chain (GVC)

In development studies, the concept of value chain has been used to analyze international trade in a global value chain, which includes "the wide range of activities required to move a product from its conception, through its design, raw materials and intermediate inputs, its marketing, its distribution and its support to the final consumer". In particular, when activities need to be coordinated across different geographies, the concept of global value chain (GVC) is used in the development literature.

Simply put, a global value chain includes all the people and activities involved in the production of a good or service and its supply chain activities at the global level, distribution, and after-sales service (also known as the supply chain) (Mehrmanesh et al., 2013). It is located at the global level. GVC is similar to the industry-level value chain but involves operations at the global level.

By joining a global value chain (GVC), small and medium-sized enterprises (SMEs) can contribute to the development of emerging economies. There is evidence that countries and companies grow and develop faster when integrated into GVCs. Participation in GVCs is uneven among emerging SMEs. Lack of access to financial services is one of the main reasons. Although SMEs account for a large share of employment and GDP, they are often unable to access much financial credit.

Sustainability standards are increasingly governing global value chains. Some of these standards are voluntary, while others are mandatory. Examples include national voluntary environmental and sustainability regulations and platforms such as fair trade or organic certification.

Leading firms often expect suppliers to comply with labor and environmental standards before they join their value chain. Some evidence suggests that compliance with these standards makes companies more profitable. Workers and the environment may also benefit. However, standards also impose costs, and as discussed in this case, standards ultimately benefit themselves.

The lack of financial resources for SMEs is one of the main reasons why SMEs struggle to meet these costs:

Governments, lenders, and businesses provide financial and technical assistance to SMEs to improve their sustainability performance.

- Development finance institutions provide bank credit to emerging countries, so they can lend to local SMEs. Some development finance institutions provide loan guarantees to developing countries so that they can lend to SMEs.
- Investors and lenders provide loans or equity directly to SMEs. Non-profit organizations provide repayable finance to SMEs to improve their sustainability performance. Some commercial banks screen sustainability applicants and offer them better terms in return for better performance.
- Some large global buyers, such as companies that lead GVCs, help suppliers access finance. Buyers also monitor suppliers' compliance with labor and environmental regulations.
- A select group of fashion retailers offers better terms in the supply chain to suppliers that comply with sustainability regulations. Some retailers offer this investment as an incentive to their suppliers to equip their customers with more energy-efficient equipment and improve labor standards.
- NGOs work with companies and suppliers to improve supply chains. For example, some NGOs suggest ways to make factories more efficient, while others help farmers use pesticides in accordance with international regulations.

- There is a growing trend for large factories to source from factories that are traditionally owned by disadvantaged groups, such as women and adults with disabilities.

The world faces a problem of integrating sustainable business practices. Sustainability data is not fully used for decision-making. To solve this problem, organizations need information systems to facilitate their sustainable actions. The Sustainable Enterprise Resource Planning (S-ERP) system is proposed as a comprehensive solution to support sustainability initiatives. The life cycle of S-ERP system is determined to include six key areas: concept, system design, initial implementation, implementation, post-implementation, and system development/retirement. A large number of research topics describe the extensive research based on key areas. The S-ERP system seems to be a promising solution to solve the problems of integration in sustainability.

8.2.3 S-ERP Metrics

As we explained, S-ERP or enterprise resource planning system has different characteristics from conventional ERP. To demonstrate these characteristics and measure the sustainability of S-ERP, we need to introduce metrics. Some of the metrics that can be used to measure the sustainability of processes and workflows of companies are as follows:

Financial sector: The financial sector is one of the most important factors in advancing an organization toward sustainability. Because companies committed to communities in this type of screening are in a better position to attract capital. These companies use three basic social commitment strategies for better financial management:

- Screening
- Support for shareholders and
- Community investment

Environmental performance measurement:

The environmental performance indicators of organizations are usually divided into three main groups:

- Environmental impact: Among them, pollution, polluting waste, energy consumption, etc. can be mentioned.
- Legal compliance: Include the status of non-compliance, costs of violating the law, number of audits, etc.
- Organizational processes: Environmental accounting, auditing, reporting, and environmental management system.

8.2.4 Sustainable Asset Management (SAM)

SAM is an international investment firm with a special focus on sustainable investing. The firm is based in Zurich, Switzerland, and incorporates economic, environmental, and social indicators into its investment strategies. In addition to asset management, the firm has indices and private equity investments. SAM has a portfolio of products based on the following areas: water, energy, climate, agriculture, and healthy living.

8.2.4.1 Cloud Computing

Cloud computing has a structure similar to a cloud cluster through which users can access applications from anywhere in the world. Therefore, cloud computing can be considered a new way to dynamically create a new generation of data centers with the help of networked virtual machines. Thus, the computing world is rapidly moving toward the development of software that is available as a service to millions of users instead of running on individual computers. Cloud computing refers to distributed and parallel systems that include a set of virtual computers that are connected to each other. These computers are dynamically provisioned and seen as one or more integrated computing resources based on service level agreements, and these agreements are established during negotiations between service providers and consumers. Cloud computing tries to dynamically create a new generation of data centers by providing services in dynamically networked virtual machines in a way that allows users to access applications from anywhere in the world.

Thus, the world of computing is rapidly moving toward developing software that, instead of running on individual computers, is available as a service to millions of consumers. It seems that selecting appropriate services from the service pool, overcoming service composition limitations, determining the importance of different qualities of service parameters, focusing on dynamic features of problems, and paying attention to rapid changes in service and network characteristics are among the most important issues that need to be examined.

8.2.5 Advantages of Cloud Computing

- Enhanced efficiency
- Lower software costs
- Fast and continuous software upgrades
- Greater document format compatibility
- Unlimited storage capacity

8.2.6 Small and Medium-Sized Enterprises (SMEs)

Small and medium-sized enterprises (SMEs) are faced with large industries and have many advantages over them, including added value, innovation, employment creation, and greater flexibility. The definition of small and medium-sized industries differs in different countries and depends on the prevailing economic and industrial conditions. Some of the criteria used to determine the type of industries (small, medium, and large) include the number of employees, capital, total assets, sales volume, and production capacity. Among them, the most common criterion is the number of employees, which is also determined differently from country to country. In Iran, there is no single definition of small and medium-sized enterprises, and it varies from organization to organization. In a classification of Iranian industrial enterprises, they are divided into four groups according to the size of the enterprise: 10–49 employees, 99–50 employees, 149–100 employees, and 150 employees and more, of which the first three groups are considered small and medium-sized enterprises.

8.3 BACKGROUND

Research activities carried out in this direction are as follows. The current service composition is difficult to achieve user requirements, because user demands change dynamically. Research, depending on user preference or environment, has created an intelligent infrastructure that can perform web service planning and composition according to dynamic user constraints. To realize this intelligent system, in this research, the exception pattern for managing user constraints that can occur during planning by creating a scenario was analyzed, and then important ideas for exception detection and improvement were extracted. Also, the system architecture was considered using exception patterns and state computation theory in first-language logic to conclude the planning method, and using this intelligent infrastructure, the complex scheme was easily designed.

To implement a smart business or provide an efficient web service, a web service composition model is required. Web service composition refers to the process of heterogeneous web service collaboration. In past works, a queuing network model has been used to evaluate the performance of composite web services. Thus, web services can be executed in an optimal manner. This model includes the essential components of a reliable web service composition. These models are so flexible that the performance indicators of composite web services can be evaluated simply by the MVA algorithm instead of performing many tedious experiments. Research shows that the model in this study provides an effective way to provide optimal service (Fazlollahtabar, 2021b).

In the literature, an attempt has been made to create a more comprehensive view of the differences in ERP implementation processes to create a sustainability perspective in processes. The first contribution of this research is to identify important differences in the key and main success factors of KCSFs between the manufacturing and service sectors. There is considerable research in identifying KCSFs, but without identifying the difference between manufacturing and service. The second contribution of this research is to differentiate ERP processes from implementation in these two sectors and, more importantly, to reject the notion that the implementation method is universal for all industries.

Cloud computing enables many web service applications to provide ERP services over the Internet. This technology has the potential to revolutionize the way IT services are used. Recent research shows that ERP delivered by SaaS has better performance than traditional IT services. However, the distribution of a service is more complex than the distribution of a product due to the impossibility of the integration principle in relation to services. In another research, a cloud-based ERP platform is defined, where enterprise customers can choose web services and customize a unique ERP system to meet their specific needs. Cloud-based ERP aims to provide enterprise users with the flexibility to rent a comprehensive ERP service from multiple vendors. This chapter also addresses the challenges of writing web services and provides a web-based solution to automate the ERP service customization process. The proposed service composition method is based on the concept of genetic algorithm and combines the knowledge of web services extracted from the web service platform with rough set theory.

Cloud-based manufacturing, or CMfg, is a new generation of service-oriented network-based manufacturing model that provides resources, capabilities, and services to distributed users in a centralized manner. CMfg is applied in the research for a semiconductor manufacturing plant. The benefits are categorized into five aspects: cost savings, efficiency, the ability to analyze additional data, flexibility, and closer partnership relationships. A SWOT analysis is performed to help the manufacturer plan the implementation of the CMfg project. A water production plant is simulated as an example. Many CMfg services are proposed to help simulate a water production plant through collaboration between cloud service providers, software vendors, equipment suppliers, etc. The relationship between sustainability and competitive advantage is also emphasized.

Resource planning of distributed organizations can be customized according to the specific requirements of each sector in network structures to enhance the value chain of SMEs. Since the implementation and control processes of distributed ERP systems in companies with limited resources are challenging, a multi-stage optimization method was proposed in past works in the evaluation of ERP systems in network companies in the food industry. The proposed method is based on the Food ERP Optimization and Orchestration Control in Network Companies (FEo2NE), and SMEs within the network can interact with services according to the optimized need using the module controller. The significant points of the capabilities of this method are the multi-criteria optimization of ERP and SME collaborations in the value chain distribution platform. According to this research, the sustainability of the SME value chain leads to cleaner production (Taherikhonakdar and Fazlollahtabar, 2024).

Identifying the different objectives of studies in the field of composite cloud service implementation and purposeful classifications of such approaches and divergence mechanisms are among the major research achievements carried out.

Service composition techniques were first introduced in a study where researchers designed a computational model that abstractly maps a workflow into a coherent workflow that satisfies user requirements with respect to quality of service parameters.

Various studies have been conducted to obtain an optimal composite service. A SaaS-based composite model was presented that considered QoS for both the service and the cloud environment. Also, a web service composite approach based on genetic algorithms was presented for web services that minimize service level agreement violations (Fazlollahtabar, 2020).

Researchers proposed using genetic programming to compose web services, which examined three types of changes to ensure that the solution is correct in terms of features and is also optimal based on the desired quality of service.

Automatic composition and workflow optimization is a challenging area of research. Several problems still need to be solved, such as how to achieve automatic architecture in such an environment where services compete with each other. In research, researchers used an intelligent approach based on the blackboard system to produce workflows in a service-rich environment, such as the grid and cloud computing.

The development of cloud computing technology creates a global market for the development and sale of products and services based on it. The increasing spread of various platforms, such as tablets and smartphones that interact with cloud computing, has accelerated the spread of this technology. Research suggests that adopting cloud-based IT services may be particularly suitable for SMEs as they have the potential to reduce some of the traditional disadvantages that SMEs have compared to large companies, such as capacity and financial constraints, and provide them with advantages that are explored in the study.

Also, with the increase in SaaS ERP solutions on the market, SMEs are considering implementing these solutions in their organizations. Another study showed that while the ERP deployment model plays a key role in enabling organizations to reduce costs, other key factors as well as barriers need to be considered. This study identifies common problems experienced in implementing and also after implementing SaaS ERP in SMEs.

Also, in another study, recent approaches to cloud manufacturing are discussed from the perspective of a sustainable manufacturer. In particular, the major differences between cloud manufacturing and web-based manufacturing are discussed. This research is based on the belief that cloud manufacturing can be a powerful supporter in the manufacturing industry, especially in a sustainable way.

In the discussion of finding a roadmap for implementing sustainable ERP, the latest research was conducted in 2016. Researchers discussed various studies regarding different proposals for a roadmap to achieve sustainability. In this research, the authors formulated a new concept based on the literature to create a roadmap for implementing sustainable ERP.

Also, in a study in 2010, a brief summary of the trends and concepts in the development of sustainable products, processes, and systems was provided. They have presented, in particular, recent trends in the development of sustainability scoring methods for products and processes and optimization techniques for sustainable production processes.

On corporate sustainability, researchers introduced a roadmap for implementing sustainability, which was formed by analyzing the business processes of an Italian company. This roadmap, called the Green Roadmap, consists of three main layers of the organization, including organization, processes, and products. The authors have argued that transforming a business into a sustainable one should start with a cultural change.

Researchers provided recommendations for the selection, implementation, and operation of sustainable ERP systems. Their study focuses on the people responsible for the ERP system, especially project managers and IT managers. Their research structure corresponds to the three stages of the life cycle of an ERP system in an organization, which include selection, implementation, and execution operations. A general process model for selection and implementation is presented. The execution operations of an ERP system are defined based on the concept of Aris.

Service-oriented systems have been considered due to their possibility of use in distributed environments. Web services are also one of the methods for implementing the concept of service orientation. In research, the issue of solving user needs by combining services has been addressed. This research has shown that, in fact, the user needs to select basic services for combination in such a way that the combination operation is more convenient and the composite service is available faster and without errors.

Software companies are looking for new and sustainable business models. The key strategy to achieve this goal is innovation. The main problem in a sector that is mainly composed of SMEs is to keep innovation a sustainable practice. Many companies have changed their system architecture to SOA, but despite its potential, SOA projects are often costly, complex, and risky. One approach to address this issue is to empower companies to innovate collaboratively. Current innovation models are largely dedicated to the manufacturing sector, which does not support many of the software and features of SOA. Past researches proposed a collaborative innovation process model to fill this gap.

8.4 MODELING THE NETWORK OF SMALL AND MEDIUM ENTERPRISES

A network of firms means firms that cooperate with each other to achieve a specific goal. Its characteristics are horizontal relationships, vertical relationships, and mutual interaction.

This concept should not be confused with personal relationships, communication activities, business connections, and cooperation in organizations. Firms in a network can reduce their costs by sharing resources and information, creating joint procurement, and reducing administrative work and bureaucracy. In addition, they can eliminate the risks associated with unknown markets and uncertainty. Four types of networks have been identified:

Cluster network: In a cluster network, firms are located close to each other. It includes suppliers, service providers, industrial firms, and institutions specialized in a certain field. These firms are competitors with each other but are willing to cooperate with each other.

Industrial region: A geographical and cultural area where employers and workers work. These firms are competitors but are willing to cooperate with each other.

Strategic network: A purposeful, long-term network between distinct organizations that is driven by the goal of gaining a competitive advantage over firms outside the network. Firms are willing to cooperate and exchange information.

Regional strategic network: A network supported by external agents and focused on firms located in a specific geographic area.

In the strategic network type, firms choose to participate in networks, which often have one manager and more than two firms. For example, in an energy-efficiency network, firms are in a group coordinated by an external agent who is constantly working to share knowledge and share experiences about energy. The coordinating element manages administrative functions and directs the network. Firms consult external experts to gain knowledge on a specific topic (e.g., energy efficiency in cross-cutting technologies, engine efficiency, etc.). After that, a common goal is defined, and the firms collaborate to achieve it. Performance in the network and firms is continuously monitored. This type of collaboration can help reduce transaction costs, minimize risks, and increase awareness about energy efficiency.

8.4.1 Characteristics of Firm Alliances

Alliance characteristics are shaped by various characteristics and influences. Although the starting point is the alliance of firms back to 1993, the model shows the scope of the cooperation structure, and therefore the model may be interesting for studying strategic networks.

8.4.1.1 Graphic Models of Networks of SMEs

- Marshallian–Italian network: A set of SMEs that all provide each other with information and raw materials. Examples include the Italian companies GOLD District Valenza and Shannon Soft.
- Multi-stage supply chain: Products are distributed in a network of different SMEs to cover different market needs (e.g., shoes for men, women, and children). This chain is therefore formed so that companies work in parallel in several stages. An example of a network type is found in the Italian shoe manufacturing sector.
- Hub and satellite network: In this network, one company acts as a leader and influences the decision-making of other companies. An example of this is the eyewear industry in Italy. Five leading companies among 1500 other SMEs are responsible for supplying products for components of other companies or for exclusive production in a specific field.
- Science Park: In addition to the three graph models described, there is another model that is more commonly used in high-tech manufacturing or service provision, which is called the science park or in France, the "pole of competition".

The lighter nodes are the services that connect the firms in the science park. The communication in this model is very flexible and informal. The network model we are considering is a cluster network for collaboration. In the SME network model, what is considered as network nodes are the SME companies, each of which has an S-ERP with the following structure, which is taken from Porter's supply chain model.

The rapid growth of technology within firms and the increasing costs of technological development have led firms to become specialists in one area to meet the requirements of their customers, with fewer skills in all areas. This has increased interdependence among SMEs and has led to a greater interest in networking by businesspeople. In order to succeed, it is claimed that SMEs need networks that are composed of various relationships. These networks consist of diverse actors, including suppliers, contractors, customers, and key users, as well as competitors, universities, R&D partners, distributors, service providers, and investment partners. Since the aim of this research is to achieve a sustainable approach with the help of information technology, the sustainability of the value chain can be achieved by collaborating services according to optimization requirements. In the following, we demonstrate the use of cloud computing technology in the design of the desired services.

8.4.2 Introduction to Layers

The layered architecture of the proposed cloud manufacturing system includes the physical resource layer, the virtual resource layer, the core service layer, the application interface layer, and the application architecture. In fact, the purpose of cloud manufacturing is to collect manufacturing resources and manufacturing capacity for centralized management. However, all kinds of manufacturing resources and manufacturing capacity are distributed across different types in different companies. Therefore, from the perspective of system engineering, the CMS cloud manufacturing system can be divided into the cloud terminal subsystem and the cloud platform subsystem according to the available space of manufacturing resources and manufacturing capacity.

The cloud terminal subsystem is basically based on the enterprise information system. It uses information technologies such as mobile Internet and Internet of Things to collect various elements of enterprise information and provide data interfaces to the cloud platform. The cloud terminal is a more detailed view of the integration of internal enterprise information, and includes systems such as product life cycle management and data management implemented by the enterprise ERP.

The cloud platform subsystem is a large data management center responsible for the convergence, integration, distribution, monitoring, and operation management of information related to production resources and production information provided by enterprises. And it provides all kinds of data interfaces and human–computer interface environments.

In addition to the layers studied, it also includes two layers of service composition and stability features. The process of service provision, request, and connection can be arranged in such a way that service providers introduce their services to the web service available in the cloud environment to display them to user requests. At the same time, users also send their service requests to the broker, which must select the best service or set of services based on the user's needs and preferences. Service providers connect related services to users according to predefined rules and agreements.

The increase in the number of available services increases the number of similar services for different services. These similar services are located in different locations and have different values in terms of service quality parameters for sustainability. For this reason, the cloud computing composite service applies appropriate techniques to select an optimal service among different similar services located on distinct servers to achieve the highest quality parameter according to user needs and user preferences. Considering the changes in cloud environments, existing services, and end-user needs, the composite service should be designed dynamically and with automated capabilities.

8.4.3 Cloud Production Subsystems

Service candidates designed to achieve sustainability: A variety of composite services can be created as candidates by combining different services provided by different

SMEs with different focuses on different aspects of the industry and collaborating with each other.

$$\text{Service} = \{T1, T2, \ldots, Tn\}$$

- Candidate #1 Service Combination
  Providing raw materials from green factories and suppliers and ensuring continuous annual communication with them and two-way reporting of the green production process
- Candidate #2 Service Combination
  In order to save energy consumption and produce an annual consumption report
- Candidate #3 Service Combination
  Using a clean and green transportation system and choosing the best time of movement and the best distance traveled
- Candidate #4 Service Combination
  Following up and ensuring continuous communication with customers, establishing after-sales service, and getting feedback from customers
- Candidate #5 Service Combination
  Implementing equal rights and benefits among employees, hiring minorities, ensuring the granting of bonuses and benefits, and paying insurance premiums to all employees
- Candidate #5 Service Combination
  Managing waste with recycling and reuse approaches and extending the responsibility of the organization during the life of the product/service, its impacts, and reuse and disposal

8.4.3.1 Service Evaluation Factors

Six QoS characteristics for service resources that the same are different SMEs defined in this work. These QoS properties are emphasized in the works of Tao et al. as a quality assessment criterion and are as follows: Time (service execution time), $T = (T1) + (T2) + (T3)$, measures the delay between the time a service request is sent and the moment the results are received. T1 is the service execution time, which measures the expected delay between the moment a task starts to execute and the moment the task is completed. T2 is the time spent by the network and intermediate hardware, and T3 is the time to transfer the service or information, estimated based on past executions of organizations. Cost (service execution cost): The price of executing (or cost) a resource service is the cost that the user requesting the service or resource pays for requesting the resource service. The service provider either advertises the price of the service when it is published or registered or provides tools for potential requesters to check on it (Taheri et al., 2014).

- Reliability: Reliability is the ability to perform under specific conditions and times, for example, the reachability rate over time, the average failure rate without failure, and the average return rate. The exact meaning and measurement of reliability vary according to the practical conditions and circumstances.

- Maintenance: Maintenance is the ability of a resource service to cope with unexpected conditions, such as network bandwidth reduction and network disconnection.
- Trust-QoS: Trust-QoS is the belief or trust that a relying entity (e.g., user or RSD) holds in a trusted domain (e.g., Requested Service Provider or RSP) for a specific period of time that gives it the confidence to perform a transaction with the trusted entity. Trust-QoS is realized through a QoS trust relationship.
- Performance similarity: Performance similarity is the probability that the performance of a requested resource service can be compared with other resource services that belong to the same service set.

8.4.4 Integration of SME Collaboration on the Cloud Platform

From a management perspective, issues such as the level of support from the organization's top management for the project are of great importance. Implementing ERP on a network scale of networked small and medium-sized enterprises requires integration across all organizational units. This requires comprehensive support from different departments and their willingness to share their data. The level of familiarity of organization managers with information analysis tools is another criterion by which the organization's ability in such projects can be measured.

The rapid development in the use of cloud computing has led to the dissemination of more cloud services in the global service pool. Due to the existence of complex and diverse services, a single simple service cannot meet the existing functional needs for many real business cases. To complete a complex service, it is necessary to have a set of simple services that work together. Therefore, there is a serious need for cloud service architecture in cloud computing. Infrastructure refers to the underlying physical components that are required for a system to perform its functions. In information systems, these components can include processors, storage, network equipment, and, in some cases, database management systems and operating systems.

A software or application running on the vendor's infrastructure is provided as a service to which the consumer has limited access. The service is provided through a thin client (e.g., a web browser) or an application programming interface to send information and receive results. The consumer is unaware of the software provider's infrastructure and has limited ability to configure some settings.

In this environment, SMEs share their data or others' data. Each of the business processing modules provided by the cloud web services helps SMEs in different parts of the supply chain.

8.4.4.1 Shared Engines in the Cloud Manufacturing Service Platform

Based on the platform system design requirements, we focus on building an intelligent logic engine for supply and demand, a logic engine for exchange coordination, and a reputation evaluation engine. These three engines reflect the core idea of cloud manufacturing services and directly decide whether this cloud manufacturing service management mode is suitable for SMEs to supply new products and services.

8.4.4.2 Intelligent Matching Engine for Supply and Demand

The intelligent matching engine for supply and demand will coordinate and optimize the mass information of service resources, share and cooperate internal resources, and share production resources as well as production capacity to realize intelligent service planning and service configuration management based on knowledge. The intelligent matching engine is mainly based on an efficient search algorithm to return the function, data, and service related to the input conditions for users. Using this intelligent matching engine, users will match the results with their input conditions to get the best service. The system mainly relies on semantic search based on keywords to find all kinds of resources. It not only supports traditional keyword search, but also supports semantic search, through which we can access data with the same or similar meaning, and overcome the limitation of concept similarity.

8.4.4.3 Transaction Coordination Logic Engine

In order to enable users to easily manage their orders and realize the communication and coordination of tasks in the transaction in the cloud production service platform, we have considered a transaction coordination logic engine in the design of the cloud platform. This engine includes three main tasks of business process management, transaction task guidance, and order monitoring. For different enterprise companies, the duration of the business process is different. The business process management tool is designed for this purpose, so that users can create or modify their own business processes in the platform according to their own business needs.

8.4.4.4 Credit Assessment Engine

The operational capacity of the enterprise users is estimated through assessment management. In this regard, the initial credit rating is the basic evaluator of credit assessment. The initial credit rating of enterprise users is based on the information provided by the users, and the system algorithm calculates it when the user registers in the environment.

An algorithm of the enterprise credit assessment engine in the cloud environment:

```
function processForm(form) {
  var xf, hc1, hc2, hc3, hc4, hc5;
  xf = hc1 = hc2 =hc3 =hc4 =hc5 = 0;
  if (form.quality[0].checked == 1)hc1 = 1;
  if (form.quality[1].checked == 1)hc1 = 2;
  .........
  if (form.service[0].checked == 1)hc2 = 1;
  if (form.service[1].checked == 1)hc2 = 2;
  .........
```

```
if (form.deliveryspeed[0].checked == 1)hc3 = 1;
if (form.deliveryspeed[1].checked == 1)hc3 = 2;
.........
if (form.attitude[0].checked == 1)hc4 = 1;
```

8.5 IMPLEMENTATION AND DATA ANALYSIS

In order to answer the research questions, the collected data must be analyzed. Therefore, in this part of the research, the collected data are analyzed based on the analysis and testing methods explained in Section 8.3. This part first presents the descriptive findings and then presents the inferential findings. Data analysis is a multi-step process in which the data obtained through the use of graphs are summarized, coded, categorized, and finally processed to provide the basis for various types of analysis and relationships between these data in order to test hypotheses; in this process, the data are refined both conceptually and empirically; Various statistical techniques play a significant role in inferences and generalizations.

8.5.1 List of Terms

CMCSS: Candidate Cloud Service Set for Manufacturing
MCS: Cloud Service for Manufacturing
CCMS: Composite Cloud Service for Manufacturing
CMfg: Cloud Manufacturing

8.5.2 Relationship of Roles in the Network

Definitions of cloud manufacturing usually refer to three clear groups of actors: consumers, who request and use cloud manufacturing processes; software providers, who enable the software to deliver cloud manufacturing and its associated ICT; and service providers who provide, own, and manage manufacturing services. Standardized descriptions of products, processes, tools, and other items are used to match the capabilities of service providers, and product requirements are mapped to a temporal supply chain. While consumer requirements include product specifications, they can also include quality specifications, cost, speed of delivery, or specific organizational requirements such as the greenness of the product manufacturing process or its environmental friendliness.

8.5.3 SME Data Flow Diagram

In the task transaction management process, we need to track and maintain the order status information in real time to realize the cooperation between the supply and demand sides. Only those workflows that remain incomplete are managed by the coordination logic engine. During the transaction task guidance process, the operation authorization of both sides changes with the change of the workflow status, as shown in Figure 8.1.

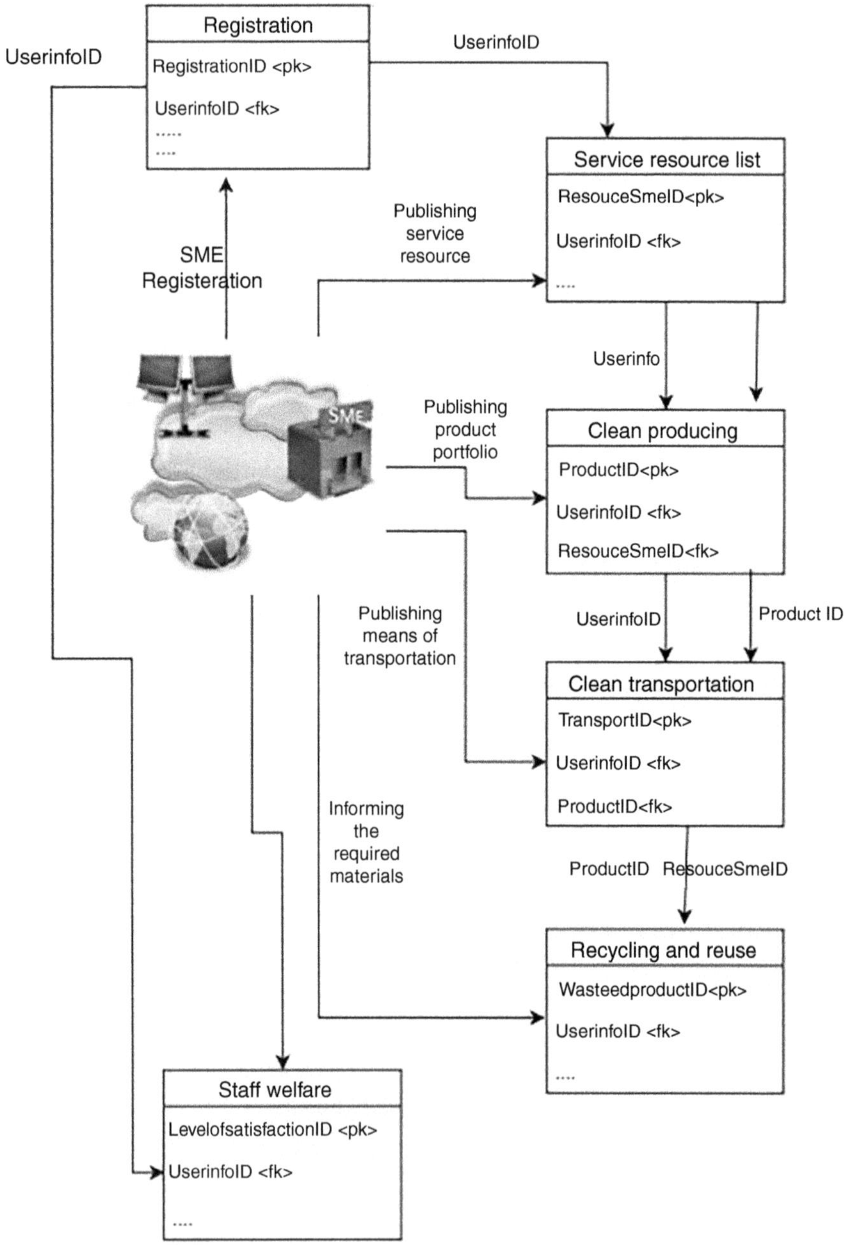

FIGURE 8.1 Data flow diagram.

With regard to the execution state on the demand side, the system sends a message to the opposite factory to enter its service status. This information rewrites the service status; at the same time, the operation authorization goes to the supplier side. This system has realized a transaction task collaboration management platform, through the operation information to the service status, in turn, from the other side.

8.5.4 Service Analysis

The process of matching the demands of service requesters and the resources of service providers should be considered from the very beginning, after the tasks are provided. In general, this process can be divided into the following three steps:

Task decomposition: First of all, the submitted task or tasks, which can be a complex production project, should be decomposed into several subtasks according to the complexity of the production processes: T= {ST1, ST2, …, STi, …, STI}, where I is the total number of subtasks that make up the task T.

Service discovery and matching: Then, for the subtask STi, and based on the operational requirements, a number of services are retrieved using a matching algorithm. This algorithm is mostly based on estimating the semantic similarity between the subtask descriptions and the resources. After this process, each subtask can be associated with any set of CMCSS cloud production service candidates.

Service aggregation and their combination: Now it is time to identify a service combination, which can be defined as selecting a candidate service from each of the corresponding candidate sets to combine the services while ensuring that the overall QoS is desirable.

In general, there may be many services with different QoS that can benefit from the performance of each subprocess. If we assume that mi is a candidate service for the subtask STi with n subtasks for the decomposed task, then the possible available services are $\prod_{i=1}^{n} mi$ for processing the task.

Where each result represents a possible composite solution for the decomposed subtasks. A common problem of service aggregation or composition is to choose the optimal composition for the execution of the task T while the desired characteristics, which are here the sustainability of the enterprise business process (time, cost, reliability, availability, and greenness), and the execution of the task is under the necessary constraints, so that the maximum QoS can be achieved.

8.5.5 Detailed Analysis of the Machinery Manufacturing Industry

From the perspective of the definition and philosophy of cloud manufacturing, the scope of manufacturing is not only limited to mechanical manufacturing but also includes machinery, electronics, chemicals, light, and food industries, and other small and medium-sized manufacturing industries. According to the standard in Guo's research, the manufacturing industry mostly includes the second industry sector, which shows more than 30 types of categories.

8.6 CASE STUDY OF GREEN MECHANICAL PARTS MANUFACTURING COMPANY

If we assume that we intend to develop a cloud manufacturing platform for machinery processing, the definition of its requirements includes the following:

(1) A third-party independent cloud platform
(2) Manufacturing machinery for general machinery parts

(3) Manufacturing processing services
(4) Network of regional manufacturing enterprises

According to the above design method and the characteristics of the mechanical processing field, a cloud manufacturing platform for machinery manufacturing processing is constructed, with a simplified construction process.

First of all, a third-party independent cloud platform should be established based on the purpose of the system. According to the business model of cloud manufacturing, which is a science park network, the independence of the cloud platform is realized. The requester and the service provider can be dynamically changed.

The operation process of the cloud manufacturing service platform is presented. In the first step, service providers can input their production information and production capacity information into the platform. They can also develop a personal data interface and place the production resource data into the resource platform database. The platform database will perform the virtualization management of the classification. After other enterprises register on the platform, they can also publish tasks related to the platform through the task publication page and manage them with the corresponding assistant. All users can also perform tasks regularly through the personal data interface.

According to the time of resource discovery by the customer, the platform can execute resource discovery. When the system finds the right resources, it will give them to the service provider. Then, the service provider decides whether to accept the job or not. According to the conditions decided by a number of service providers, the system forms a set of service provider candidates and waits for the customer's decision-making.

When the customer and the service provider finally establish a cooperation agreement, the system starts to assign tasks. Then, the service providers carry out the work according to the contract and coordinate the quality of production services within a certain period of time. After completing all the tasks, the service notifies the cloud provider and distributes the relevant facilities to the platform for implementation. When customers receive the product, they will pay under the support of the system. Both parties will conduct a mutual evaluation, and then the system will terminate this production service.

8.6.1 A Study on the Efficiency of Composite Service in the Automotive and Industrial Parts Manufacturing Industry

In this section, we introduce the identification of processes, services, and specifications, as well as the desired artifacts obtained from the studied industry of machine parts, and propose a model that takes into account the optimal composite service. Finally, the results of the studies on the artifacts produced by automobile manufacturers and a brief summary of the research results are presented.

8.6.2 Small and Medium Enterprises Manufacturing Automotive Mechanical Parts

In this regard, in order to achieve the physical and basic layer of the model, which includes a network of small and medium enterprises cooperating with a common strategy, we introduce the hub network set in the real market.

8.6.3 Set of Automotive Parts Manufacturing Companies

Marpich Bakhtar Company
Samim Behrad Company
Gardoon Gas Jack Industries Company
Bekaran Sanat Helia
Takvin Electric
Azar Batterie Urmia Company
Aria Plast Azin Company
Asia Lent Company
Azbar Andisheh Company
MP Sport Industrial Group

8.6.4 Service, Service Identification, and Service Details

Design principles should also be considered when identifying and specifying the services of the parts manufacturing industry, as given in Table 8.1

Service specifications are essential for identifying and reusing a service. A service specification defines a service according to predefined principles. It includes all the information required by various stakeholders for design, development, maintenance, and use.

Service Detail Groups: While a technical service specification covers implementation-related aspects, a service concept specification focuses more on information related to the context in which the service is used and organizational aspects.

Group-level description: Operational specifications of service behavior (semantics, data meanings) in a process context and specifications of pre- and post-conditions for correct service execution. Example: Service operations, information objects used, input/output

TABLE 8.1
Service Design Principles

Design Principles	Description
Interface Familiarization	Comprehensive standardization of service details for component manufacturing Sustainable service contracts
Interoperability	Technical and conceptual interface standardization
Autonomy and Modularity	High flexibility and weak logical coupling Weakly coupled communications
Demand/Process Guidance	Grain details of the service with respect to the business concept/component business Generalization of service outputs
Abstraction	Platform-independent service description and implementation

8.6.5 Service Operations of Part Production

Information objects: digital images of products
Conceptual semantic input and output
Aspects related to the use and maintenance of a service

Example: Operational service name (part raw material provision, part production, testing)

Business contact person (manager of various departments such as production, sales, and logistics)
Management part version number
Information needed to ensure the technical connectivity of the service

Example: Technical service name, data formats/types, technical operational network address

Non-operational requirements and guidelines for using the service in the desired quality conditions

Example: Service cost, greenness, security (including raw material provision, production, warehousing, and transportation), quality

8.6.6 Description of Service Detail Groups

Process services: covers operational capabilities that can be directly extracted from the business process model and support process activities. They can call other process services, rule services, and entity services during runtime.
Rule services: contain business rules and validation. They are called by process services (e.g., during calculation).
Entity service: allows operations (e.g., create, read, update, delete) on an information object. These services provide a standard view of the information object, and these services allow the creation and modification of data in an information object by calling rule services and process services.

Several process services, rule services, and entity services can form a clustered service network based on their functional and logical similarity.

8.6.7 Service Design

The following composite service map is the result of the approach outlined in the previous sections. This standard map standardizes the ERP services of small and medium-sized enterprises and the services associated with specific service categories. In order to reduce complexity, the diagrams are limited to process service and entity service as the main service types.

Work Management: Work management includes all services related to personnel management, personnel accounting, personnel planning and capacity planning, personnel allocation to programs, and time and attendance reporting. In this service combination, we have emphasized the social sustainability aspect by considering employee welfare considerations.

8.6.8 General/Detailed Planning

The general planning service combination includes all services related to performing rough planning, allocating production service requirements to different production sites in different SMEs in the parts manufacturing industry, creating MRP2, preparing orders, and checking their feasibility. Detailed planning includes all services related to executing detailed planning, checking constraints, sequence planning, and creating production orders and production plans.

8.6.9 Inventory Management Service

Production inventory management includes all services related to the collection and management of raw material warehouse information, monitoring and accounting for the movement of goods, and maintaining a physical inventory list. In this model, networked companies can optimize production by exchanging information to match market needs. This information is carried out through the Internet of Things and updated in web services available on the cloud platform.

8.6.10 Equipment and Production Line Management

Equipment management includes all services related to managing and processing equipment information, providing equipment in the production process, ensuring maintenance and repair planning, responding to urgent problems, performing maintenance reports, and monitoring the correct operation of equipment in line with sustainability. In this section, companies can benefit from the cooperation of other companies by announcing the status of their equipment.

8.6.11 Quality Management

Quality management includes all services related to quality planning and inspection, executing and documenting quality inspections, providing information for production documents, creating quality analysis reports, recommending and improving actions, and determining and controlling revisions. At this stage, companies' real-time monitoring of each other on the cloud improves their performance.

8.6.12 Resource Management

Resource management includes all services related to ensuring resource availability for processing, managing, and processing the inventory of running processes, tracking resource status, maintaining resource history, changing plans if necessary, reserving resources, and dispatching resources. By using the SME network under the cloud platform in this section, comprehensive monitoring of energy consumption, equipment emissions, and the greenness of the production plan can be achieved.

8.6.13 Resource Management Service Composition

Production Analysis and Reporting Service Composition:

Production Analysis and Reporting: Production Analysis and Reporting includes all services related to minute-by-minute reporting of production processes, long-term production analysis, visualization of production data, and exception management.

8.6.14 Manufacturing Process Execution/Control Composite Service

Production monitoring and execution includes all services related to monitoring production orders, controlling and adjusting monitoring sequences, comparing planned and actual production status, updating production plans, creating efficiency of planning schedule programs, locking parts and recording scrap, notifying raw material movements, managing inventory, and sending orders for quality checks during process execution.

8.6.15 Research Questions Analysis

- How is it possible to link sustainable ERP systems with the SME sector?
  This question was examined in the third and fourth sections by showing the cooperation model of SMEs, and it was found that SMEs in a cluster network can achieve sustainability by integrating their ERP systems, input and output data, and having a database, by combining their services.
- What is the role of SMEs in sustainable development?
  Since SMEs are very flexible in the process of adaptation to continuous changes in the environment, because they are in contact with customers and provide appropriate feedback, and by using the advantages of online marketing, they continuously innovate in order not to be left out of the market competition. SMEs and their investors should be involved in the decision-making process at the global and regional levels. The Global Strategic Directive emphasizes the role of SMEs in sustainable development. In order to achieve this goal, as explained in this study, SMEs should use the network facilities of which they are a part. But finding economic arguments in favor of sustainability can be challenging. At the same time, numbers are not the only measure of success. For example, sustainable measures may maintain employee employment or social relations.
- What are the optimality indicators for a composite service architecture?

The sustainability indicators within the organization are as follows, and the following aspects are shown:

Company reputation: financial strength, number of employees, and trainees

Economic: treasury level, work results, return on investment

Suppliers: meeting the need from the manufacturing machine/raw material, service, supplier diversity

Customers: customer satisfaction, customer return, successful contracts
The employees themselves and the organization structure: presence at the workplace, skills and knowledge, information flow
Products: number of product innovations and having an environmentally friendly production process.

The optimality indicators that we presented for combining services in different parts of the supply chain in Section 8.4 include making the value chain more profitable, increasing the level of well-being of industry employees, and thus raising the social image of companies, as well as reducing energy, water, pollution, raw material consumption, the amount of recycled materials consumed, the number of transportation, and the amount of energy consumed as a result, the amount of equipment depreciation, etc.

8.6.16 Analysis of the Proposed Model Using the Peer-Review Method

Peer review is called expert review and is a well-known method for advancing the research process and research results using experienced and specialized experts. These experts will evaluate the important and consultative assessment related to the quality of the research project and the research results. Experts help to present and support them, challenge the assumptions of the study and its findings, and help the researcher improve the validity and accuracy of the study.

The peer-review process can be presented through formal, written reports or through informal conversations and emails. Experts can also improve their data analysis and interpretation by supporting peer reviewers and using their feedback to reach consensus on the consistency and agreement of the findings.

We have used the common single-blind method for this drug, in which the author does not know the names of the expert reviewers.

Advantages: Reviewer anonymity allows for unbiased decision-making – authors should not be influenced by recipients

Disadvantages: Reviewers may use their anonymity as a justification for harsh and unnecessary criticism when commenting on the author's research.

8.6.17 Suggested Form for Collecting Expert Information

In order to validate the results obtained from 10 expert reviewers on the topics of cloud computing and the sustainability of ERP composite services for SMEs on the proposed model, after they carefully read this section, we collected their opinions using the following survey form given in Table 8.2.

The rating is as follows: [1–2]: Weak; [3]: Satisfactory; [4–5]: Strong

Table 8.3 shows how many experts gave scores in each score range for each criterion.

Regarding the clarity and importance of the topic raised, the scores obtained in the 10-person set and in the range of [1–5] were as follows: Total score = {4, 4, 5, 3, 3, 3,3,3, 3, 1, 2}

TABLE 8.2
Service Composition for ERP Survey Form

Criteria	Expert Opinion	Strong	Satisfactory	Weak
Topic raised: Clarity, importance				
Evidence: Relevance, strength, credibility				
Order: Organization of ideas, reader guidance				
Optimality result				
Sustainability result				
Overall effectiveness				

TABLE 8.3
Experts' Scores for ERP Service Composition

11.719 pt	Expert Opinion	Strong	Satisfactory	Weak
Topic raised: Clarity, importance		3 people (30%)	5 people (50%)	2 people (20%)
Evidence: Relevance, strength, credibility		2 people (20%)	5 people (50%)	3 people (30%)
Order: Organization of ideas, reader guidance		4 people (40%)	4 people (40%)	2 people (20%)
Optimality result		3 people (30%)	4 people (40%)	3 people (30%)
Sustainability result		4 people (40%)	3 people (30%)	3 people (30%)
Overall effectiveness		4 people (40%)	4 people (40%)	2 people (20%)

From this total set, each score is placed into one of three subsets, sorted by rank, according to its value:

$$\text{Strong score set} = \{4, 4, 5\} \Rightarrow |\text{Strong score set}| = 3$$

$$\text{Satisfactory score set} = \{3, 3, 3, 3, 3\} \Rightarrow |\text{Satisfactory score set}| = 5$$

$$\text{Weak score set} = \{1, 2\} \Rightarrow |\text{Weak score set}| = 2$$

30% strong, 50% satisfactory, 20% weak

Regarding the validity, relevance, and strength of the evidence presented for the analysis of the model, the scores obtained in the 10-person set and in the score range [1–5] were as follows:

Overall score set = {4, 4, 3, 3, 3, 3, 1, 2, 1}

From this overall set, each score is placed in one of three subsets classified in order of rank according to its value:

Strong score set = {4, 4} => | Strong score set |= 2

Satisfactory score set = {3, 3, 3, 3} => | Satisfactory score set |= 4

Weak score set = {1, 2, 1} => | Weak score set |= 3

20% strong, 40% satisfactory, 30% weak

Regarding the idea organization criterion, the reader is guided while analyzing the model of points obtained in a set of 10 people and in the score range [1–5] as follows:

Overall score set = {4, 5, 4, 4, 3, 3, 3, 3, 2, 2}

From this overall set, each point is placed in one of three subsets classified in order of rank according to its value:

Strong score set = {4, 5, 4, 4} => | Strong score set |= 4

Satisfactory score set = {3, 3, 3, 3} => | Satisfactory score set |= 4

Weak score set = {2, 2} => | Weak score set |= 2

40% strong, 40% satisfactory, 20% weak

Regarding the optimality criterion, the result obtained from the proposed model was the scores obtained in the 10-person set and in the score range [1–5] as follows:

Overall score set = {4, 5, 4, 3, 3, 3, 3, 2, 2, 1}

From this overall set, each score is placed in one of three subsets classified in order of rank according to its value:

Strong score set = {4, 5, 4,} => | Strong score set |= 3

Satisfactory score set = {3, 3, 3, 3} => | Satisfactory score set |= 4

Weak score set = {2, 2, 1} => | Weak score set |= 3

30% strong, 40% satisfactory, 30% weak

Regarding the sustainability criterion, the results obtained from the proposed model are as follows:

Overall score set = {4, 5, 4, 4, 3, 3, 3, 3, 2, 1}

From this overall set, each score is placed in one of three subsets classified in order of rank according to its value:

$$\text{Strong score set} = \{4, 5, 4, 4\} \Rightarrow |\text{ Strong score set }| = 4$$

$$\text{Satisfactory score set} = \{3, 3, 3, 3\} \Rightarrow |\text{ Satisfactory score set }| = 4$$

$$\text{Weak score set} = \{2, 1\} \Rightarrow |\text{ Weak Score Set }| = 2$$

40% Strong, 40% Satisfactory, 20% Weak

According to the results obtained, more than 70% of experts feel satisfied with the overall result of the work.

8.7 CONCLUSION AND SUGGESTIONS

The results of research work are usually read before the other parts. In light of reading these few pages of conclusions, the reader will get a general idea of the benefits of the research for them without having to read the entire report. Based on this quick recognition, they will decide whether or not to read the entire report or possibly parts of it. Therefore, it is necessary to write the results of research work carefully and present useful information for potential readers. The conclusion of a research work generally consists of three parts: first, recalling the main lines of the scientific method followed, then the new knowledge achievements that the research originated from, and finally, providing practical suggestions. In this section, the results of the present research are presented, taking into account the data analysis and the results of the tests carried out.

In order to move toward sustainability, companies must simultaneously improve the social, environmental, and economic performance of their businesses. The implementation of the S-ERP system is required to integrate each sustainability process, information, and data along the advanced value chain. The aim of this study was to present a service composition model to obtain S-ERP systems considering sustainability factors, which was displayed with a service diagram.

The cloud manufacturing service platform for SMEs collaboration in a cluster network was designed as a platform that provides composite services of different processes, including production, transaction management, and employee tasks, using web services. This network modeling of small and medium-sized manufacturers has created production collaboration among companies by combining their capacities and capabilities in every aspect. This model also suggests that the S-ERP of companies should be equipped with intelligent engines to match the service requirements of each company, validate their services, and coordinate their transactions in the green supply chain. There is no doubt that the cloud manufacturing service platform is attracting attention in the manufacturing industry due to its convenience, simplicity, and powerful performance. It integrates mass production plants from SMEs and combines the advantages of each company, ultimately helping to improve the competitiveness for survival of SMEs.

The present research has played an important role in increasing the sustainability of companies as an important competitive advantage in the current era by relying on new technologies such as cloud computing. The use of other technologies, such as Big Data and IoT, to increase the level of collaboration of networked companies and the level of system redundancy by creating data input and output terminals at more nodes of each part of the value chain has achieved greater efficiency. Also, future research can be conducted in order to compare S-ERP systems with previous related systems. Since S-ERP appears as a new system, research on the previous system will be useful to determine the direction of future research and system development. This comparative research identifies and analyzes the similarities and differences in concept, benefits, issues, challenges, and implementation between ERP and S-ERP systems.

9 Business Intelligence (BI) for Insurance System Using Social Network Analysis

9.1 INTRODUCTION

By examining the speed of growth of information technology in all daily affairs of today's life, we realize that information technology and its many branches are intertwined with the lives of individuals and all organizations, offices, companies, and all business, welfare, social, and political environments in society. In a way that all the industries mentioned above consider the main factor of their survival, growth, and progress in the ecosystem related to their industry to be increasing information technology knowledge, information technology can be considered a powerful tool for competing with competitors. Information technology knowledge can be considered to include hardware, software, and brainware. To the extent that IT experts believe that the 21st century can be considered the end of the industrial and software era and the beginning of the knowledge-based and brainware era, perhaps the most important concern of our society today is that due to the lack of adherence to scientific principles and rules and the lack of effective efforts to build an appropriate culture, there have been very limited applications of information and communication technology in various fields of industry, and it has not achieved its expansion and development capabilities as it should (Fazlollahtabar, 2020).

Competitive advantage in the discussion of information technology can also include strong infrastructures, including strong local area networks and strong network actives (routing and switching), or it can include strong software in databases or data warehouses or decision support systems, or it can include strong brainware by employing experts in the field of IT. Obtaining information about approaches that provide access to big data with very high levels of data processing is desirable for organizations, departments, and companies, especially large-scale government departments (Fazlollahtabar and Mohseninejad, 2015).

Next, after introducing the Iranian Social Security Organization as a very large organization that uses information technology and its branches in the context of assigned tasks in order to achieve organizational goals, including the country's largest network infrastructure, the largest group using Cisco brand routing and switching equipment, which is one of the world's leading IT companies in the country, strong network security equipment, strong databases, and decision support systems, we will proceed with the discussion.

DOI: 10.1201/9781003740148-9

The Social Security Organization, as the largest insurance organization in the country, uses numerous electronic systems such as the Saba system, the centralized insurance system, and the centralized hospital system to achieve the insurance goals of its customers. These systems can also be referred to as decision support systems, which, in addition to having strong databases of insurance customers across the country, also have computational facilities at various management levels of the Social Security Organization. For example, operations such as calculating medical entitlement payments, calculating insurance premiums for men's military service, unemployment insurance calculations, etc., are among the functions of these decision support systems at executive management levels. Given the current conditions of the country, calculating workshops eligible for compensation for damage caused by Corona and presenting it to higher authorities in the Ministry of Labor are among the functions of these decision support systems at middle management levels and deputy directors of the organization (Fazlollahtabar and Mohseninejad, 2015). At the senior management decision levels of the Social Security Organization, Saba, and centralized systems are usually less considered, and senior managers usually tend to view the graphs and charts they need in the BI system. Requests from senior managers are made in the BI system, and BI experts and programmers, with the help of experts and programmers from the central and centralized systems, obtain the required tables and data from the central and centralized systems and load them into the BI system data warehouses. Finally, the required results can be viewed in the BI management dashboards in the form of charts, graphs, and diagrams. In this research, we are trying to develop the communication module of the BI system with its subsystems, including the central and centralized systems and the integrated hospital system, in such a way that the BI system can retrieve the required information from the central and centralized systems databases with the least amount of data exchange, store it in its data warehouse, and use it for analysis and presentation of appropriate reports for managers (Fazlollahtabar, 2021b).

The roadmap or conceptual model in this research is expressed as follows, considering the existing elements and their relationship, in a way that no change is made to the model elements, and it tries to change and improve the relationships with each other. The existing relationship includes the role and ability of decision support system programmers, calling information from databases, connecting correctly to the BI data warehouse, and communicating with BI programmers. The improved relationship tries to reduce the role of decision support system programmers and the BI system. Fetching and retrieving information from decision support system databases should be done in a way that minimizes network occupancy and creates parallel information in the decision support system databases and the BI system data warehouse.

9.1.1 Objectives

- Reducing the role of BI system programmers in the connection between the BI system and the operational systems of the Social Security Organization
- Determining and creating a control code for faster access to the database
- Making the organization's operations more systematic, which results in reducing costs and increasing the speed of response

- Clustering information in data warehouse entries
- Reducing the amount of data in the BI system's data warehouse by using unique codes to directly retrieve data from the operational systems' databases

The level under the coverage of the operational programs of the Social Security Organization, which includes all branches and medical units throughout the country.

9.2 SOCIAL SECURITY ORGANIZATION OPERATIONAL SYSTEMS

The Saba Decision Support System, an abbreviation of the Iranian Insurance System, was written in 2005 by the organization's internal programmers based on the .NET programming language on the Windows operating system and was put into operation by the end of 2006. This program was installed as a desktop on the clients of the Social Security Organization's branches and had a separate database in each unit, so that information related to the Social Security Organization's application programs, including technical, income, pension, inspection, registration, booklet, executive, etc., was stored in the database of this decision support system in the branches, and the information required for higher management levels was made available to the general departments and ultimately to the organization's headquarters in the form of consolidated reports. The dispersion of information across the country and low security in maintaining the main insured data led the organization to centralize data and use a new decision support system called the centralized insurance system. (Website: Social Security Organization)

Technological changes in recent years have had a great impact on the insurance industry, one of the most important of which is the discussion of using a centralized and integrated insurance system. Centralized insurance, known in the international market as (CORE Insurance), is a system in which a set of insurance industry business processes and operations, including issuing insurance, issuing supplements, reporting accidents, creating claim files, financial management of insurance policies and claims, etc., are planned, policymaking, managed, implemented, and supported (Bagheri et al., 2018).

Centralized insurance is a comprehensive and integrated solution that has various cores, layers, engines, systems, subsystems, and modules that are installed in the insurance company's data center and accessed through communication platforms by branches and representatives of insurance companies to carry out insurance operations. Also, through APIs and web services provided through this system, various communication channels are created between the system and the customer so that, in the form of an electronic insurance infrastructure, insurance services and products can be provided to customers on web, mobile, etc. platforms without time and place restrictions. Before using the centralized insurance system, insurance companies used to install the system on the computers of their branches and agencies, which caused challenges and inadequacies such as the creation of island systems, redundancy in recording customer information, insurance policies, claims, etc., delays in preparing analytical and performance management reports, lack of integrated and centralized policymaking and supervision, etc. On the other hand, fraud and violations that arose due to the lack of centralized information on insurance policies, supplements, and claims took time to identify or were not possible to identify.

The use of a centralized and integrated insurance system, in addition to largely resolving the aforementioned problems and challenges, has made it possible to centrally monitor all insurance operations and to centrally define various policies for user access and permissions, and to monitor user events and behaviors and identify fraud in a timely manner by implementing fraud detection patterns (Fazlollahtabar et al., 2010).

Some other benefits of centralized insurance are as follows:

1. Insurance companies can provide services and products to customers online through various mobile, web, and even landline channels, at home, at work, and even while traveling, and take action to satisfy their customers.
2. Centralized insurance will make it possible to provide electronic insurance infrastructure and systems, which, in addition to satisfying employees, customers, and other stakeholders of the company, will enable insurance companies to take effective action in line with their social responsibility to reduce traffic, pollution, and the spread of the coronavirus disease that society is currently facing.
3. Information security in island and decentralized systems faced many problems, so that in addition to information disclosure, in case of forgetting to back up information by the branch or agency or a problem with the media on which the backup was made, it was not possible to restore information in cases where the hardware or software of the branch or agency system encountered a problem, and this caused information loss. The use of centralized insurance, in addition to greatly increasing information security, has also provided greater assurance of the accuracy of maintaining information archives through backup devices and mechanisms.
4. Applying centralized and integrated technical policies, for example, after changes in premium rate calculations are applied in the system in a field, all issuing units can immediately use it for their new insurance policies.
5. Applying centralized and integrated financial policies, for example, after new financial licenses and authorities are applied in the system for a group of users, a field, or a sales network, it will be immediately available in insurance operations.
6. Possibility of online exchange with the Central Insurance supervisory systems, for example, sending draft insurance policy information to receive a code and issuing an insurance policy after obtaining a code from the Central Insurance, which has prevented the redundancy of insurance policies in different companies, provided correct and complete registration of the required information, and online monitoring of the implementation of the Central Insurance regulations and standards.
7. Reducing costs and improving operations are other advantages of centralized insurance, so that, in addition to accelerating insurance operations, all

information is consolidated in one center, and processing on the company's users' computers is minimized.

8. By using centralized insurance, the insurance company can receive timely warnings of the risks it faces based on its data analysis and take action to reduce them. For example, the risk of increasing the loss ratio in a field, after identifying its factors, new technical and financial policies can be implemented in the system to reduce the risk.
9. Preparing timely management, analytical, and performance reports so that senior and middle managers can review the status of the insurance company based on them and take timely action to identify problems and deficiencies and plan to resolve them.

Currently, in Iran, various companies are working on the design, production, and support of centralized insurance systems, with a major part of the market in the hands of one company and the rest in the hands of other companies. On the other hand, companies have invested in the design and production of their own centralized insurance systems based on their own policies, some of which have been successful and others have failed.

Based on the presented architecture and structure, in order for a centralized insurance system to be agile, intelligent, and cost-effective for insurance companies, the following layers must be considered in its design:

- Operations management layer.
- System data and integration layer.
- Shared services layer.
- Business processes and core tasks layer.
- Interaction layer with service and product delivery channels.

Within each layer, there are cores, systems, subsystems, and modules that are configured and adjusted based on the technical, financial, and sales network policies of each company.

In the Social Security Organization, since 2013, there has been an attempt to centralize insurance services in a centralized insurance decision support system, and these reforms have so far completely transferred the technical, pension, enrollment, individual accounts, and booklet sections to the centralized system, and some facilities such as income and administration are still in the process of transferring from Saba to centralized. As stated in the general explanation of centralization, the centralized insurance system uses a central database in the headquarters with redundancy and security from up-to-date backups, and provides data to the BI system data warehouse at a higher speed. (Website: Social Security Organization)

9.2.1 Hospital Integrated System

In the past few years, our country's health-treatment centers, especially hospitals, have sought to mechanize their information systems. Initially, the purpose of such activities was to reduce the costs of paperwork in manual and administrative systems, but now we have reached a stage where improving the quality of medical services is becoming increasingly important. The goal is to store all of an individual's health and medical information throughout their lifetime in an electronic file that can be accessed anywhere in the country through a nationwide computer network. Today, advanced computer technology is recognized as a reliable and cost-effective tool for integrating various software applications in healthcare environments.

Unfortunately, most of the software available in Iran has been developed to meet customer demand and not based on a comprehensive, logical, and systematic approach. Therefore, since they have been developed for different customers, they are not compatible with each other, and establishing communication between them in a network environment is very difficult and costly, both financially for the buyer and in terms of engineering personnel for the manufacturer. The existence of agreed standards, by reducing the aforementioned costs, is in the interests of both the manufacturers of hospital information software and their users.

Compliance with the most comprehensive global standard for medical information exchange, called HL7, which is currently being used worldwide for 24 years. Having the most complete and strongest security system in terms of user access levels and scope.

Having and supporting most of the leading medical engineering standards.

Including the richest medical databases, including SNOMED, ICD.9, ICD.10, California tariff, etc.

Designing the system based on medical events so that it easily adapts to the current workflow of clinical environments.

Having a complete Visual environment, the ability to fully use the Mouse in entering information so that only 2.5% of information is typed.

Great flexibility and ease of the system in changing and adapting to different hospital cultures The ability to create information exchange between hospitals at different city, provincial, and national levels. The ability to create Internet and intranet connections, support and use of common and reliable systems, electronic signatures in the form of smart cards and fingerprints, and the use of software systems for archiving, storing, and retrieving visual information.

Some of the different functions of HIS are:

- Patient Management: Various stages from the time of admission to the patient's discharge.
- Paraclinic: Related to the paraclinical departments of the laboratory and radiology. Receiving all test and radiography requests and responding to them will be possible by entering this section.
- Pharmacy: Receiving drug requests and recording information related to responding to them is done in this section.
- Warehouse: For performing warehousing operations.
- Statistics and Reports: Operation of departments, paraclinic, inpatient, and outpatient visits, surgical and anesthesia procedures, with chart display.
- Accounting: Necessary steps for final settlement with the patient, issuing invoices, and allowing discharge from the hospital, as well as defining basic accounting information, are done in this section.

Insurance: For performing insurance expert affairs.
Work: Calculation of doctors' performance
Clinic: Includes all matters related to clinic clients, such as doctor visits, injections, laboratory, radiology, etc. on an outpatient basis.
Medical Assistance: Patients who request a discount on their bill, after requesting it, the accounting department will refer their request and file to the medical assistance department. After investigating, if the patient in question meets special conditions, the medical assistance department will apply the necessary discount from specific financial resources on the bill.
Medical Documents: All surgical procedures and disease diagnoses are coded based on the ICD.9, ICD.10 books.
Medical Equipment: Recording the date of production, purchase, and warranty period of the device to support other services, and, if necessary, recording other information related to the relevant company.
Employees: Including hospital staff files (electronic personnel files).
Hospital Physical Chart: Information related to the hospital structure, including departments, wards, rooms, etc.

The HIS system in the Social Security Organization was also launched in 2017 after about two years of reviewing and piloting three models of these systems in the form of web and client-server. After that, all data conversions were performed on this system as a macro decision support system at the treatment level of the Social Security Organization. (Website: Social Security Electronic Health Company)

9.3 PROBLEM DEFINITION

Large organizations that are classified based on the number of personnel and geographical area of activity have concluded that by placing the IT line in their organizational chart in all areas of activity and organizational architecture, they can survive and progress as a successful organization. Among the IT tools that help organizations achieve this goal are BI systems and other decision support systems in the organization, which are also referred to as operational systems. Recognizing the components of the BI system and its relationship with its operational subsystems, including input data in operational system databases, structured data in the data warehouse, and tools that display the output of the BI system, such as Microsoft Power BI and SQL BI, are among the needs of the problem statement in this chapter. The relationship between input data and output information, as well as reducing the volume of ETL data in the BI system, is a desirable service organization for us in this research. Then, by reviewing the organizational architecture diagrams, we will increase our knowledge of a large service organization in which we intend to improve the performance of the BI system. And finally, by introducing the social network analysis algorithm, we will try to announce our proposed method to improve the performance of the BI system in a service organization.

In this section, we will introduce a service organization as a platform for implementing BI on it to introduce the proposed method. Before examining the information structure of the organization, we will briefly mention some things about the organizational structure of a service organization. An organizational structure creates a hierarchy between the standards and functions of the organization. It expresses how we can create work groups in an organization and establish a coordinated relationship between them in order to achieve organizational goals. After creating these working groups, the organizational structure of the organization describes the positions and duties in the group according to the dimensions of the organization, regulates the rules, regulations, and policies, determines the communication points of the working groups, and ultimately achieves organizational goals.

The notable features of an organizational structure are:

1. Focus and Decentralization
   In a centralized service organization, all decisions and planning to achieve the organization's goals are made by top-level management, and the centralized organizational structure focuses on the strong connection of the system components from top to bottom to achieve the decisions of senior management at the executive levels.
   While in a decentralized structure, the decision-making power in the lower components of the organization is increased even at the operational levels.
2. Span of Control
   In any service organization, the number of people reporting to upper-level managers in the organizational structure is considered the span of control. In other words, your span of control as a manager is the number of people reporting to you.
3. Grouping
   In this process, you divide the personnel of a service organization into smaller groups or departments in an organizational structure. The general tasks of these groups are predetermined in order to achieve the overall goals of the group and are available to the managers and supervisors of these teams.
4. Expert Specialization
   This feature includes the division of labor between subordinates according to their work specialization in order to achieve the goals of the organization. The personnel of a service organization are placed in work groups according to their degrees and qualifications, backgrounds, experiences, and specializations in such a way that the efficiency of the group is at its highest level in order to achieve the job descriptions.
5. Reporting
   The chain of command in a service organization is related to the hierarchical reporting of each manager to the upper manager. Reports are mostly given to the upper manager in the form of predetermined formats and tables, taking into account the principle of conciseness and usefulness of the report information. The reviewed reports are validated, and the necessary decisions are made to continue the process, change the process, and modify the work process.

A service organization, such as a bank or an insurance company, is in the size of large organizations and, based on the number of branches or number of agencies, the number of personnel, and the volume of work they do, these types of service organizations can be structured by the category of line and staff organizational structures, functional structure, or matrix structure.

In the line and staff organizational structure, it is very important for managers to pay attention to the opinions of consultants and experts in various fields. For example, a senior manager pays attention to the opinions of deputies, administrative financial consultants, and technical consultants in order to make a strategic decision and announce it to the middle managers, then makes a decision and communicates it to the middle managers. Similarly, middle managers also use the opinions of industry experts, technical experts, and system engineers to make decisions in their area of responsibility, and then make a decision and then communicate these decisions to operational managers, and in this way, they proceed toward the goals of the organization until the decisions are implemented in the executive ranks. In this structure, advisory teams play a significant role in improving management in various fields.

In the functional structure, however, experts and personnel are divided and grouped based on their knowledge and expertise. The scope of activity of this structure is mostly in the executive and operational management sector. While the matrix structure is a combination of most existing structures that provide solutions at different levels of a large service organization, this structure shares the organization's resources according to the geographical area. Flexibility in decisions in this structure is high. After this brief digression into the organizational structure, we will examine an insurance organization specifically as an organization by delving into its details. The headquarters of an insurance company, as the headquarters of the organization, is the location of the organization's senior managers, including the CEO and board members. Strategic and macro-organizational decisions and future planning are made in this headquarters using information from the BI system, decision support systems, and operational subsystems. For example, senior managers in this complex view the output charts of the BI system in their insurance organization in the areas of insured costs and income, costs and income related to domestic and foreign investments, to make macro-financial decisions. These output charts are displayed in the BI system by tools such as Microsoft Power BI, SQL BI, SAS Visual Analytics, Locker, Sisense, etc. and obtain the information they need based on the formatted data that is loaded onto the BI system data warehouse. The formatted data stored on the BI system data warehouse can be called information in relation to the original raw data stored in the insurance subsystems database of this service organization, meaning we obtain the data available in the insurance operational systems of this service organization, format them in information tables, and load them into the BI system data warehouse of this service organization. The data stored in the database of this organization's insurance operational systems can be personal profiles of individuals covered by this service organization, in which data such as monthly premiums paid, types of insurance used, including supplementary treatment, automobile, life and accident, liability, etc., the number of times insurance claims are used, the geographical area of use of insurance claims, etc. are stored.

A number of decision support systems used in this service organization store the organization's internal data, not data related to the organization's products and services, in their databases. Data related to the organization's personnel, including official, contractual, contractor, and even project and part-time employees, or data related to investors unrelated to the insurance services provided by this service organization, such as investments in the fields of import and export, construction, etc., are stored in the databases of these decision support systems. After performing mathematical calculations on them, this raw data is loaded into the BI system data warehouse of this organization and is made available for future exploitation. In other words, we say that after performing mathematical calculations on them, the raw data is ETLed and converted into a form of information as formatted data and loaded into the BI system data warehouse of the service organization. For example, the functions of sum, average, minimum, and maximum insurance claims received by the insured in a geographical area are simple examples of information tables created in the BI system data warehouse of this organization. Services that are obtained by performing mathematical calculations on data stored in the databases of the insurance operational systems of this insurance organization.

Suppose that information is required to have the number of people using the life insurance death claim service whose age is between 50 and 70 years, first we extract the list of all insured people in a table from the organization's operational programs, then we give only the life insurance column of this table a weight of one and the weight of the other columns is zero, then we select the age range of 50 to 70 years in it, and finally in the resulting table we select only the rows that have used the life insurance death claim service, and the resulting information is stored in a final table in the data warehouse.

What we want, and what we will refer to as the proposed method in this chapter, is to use some of the capabilities of the social network analysis algorithm, such as sampling and clustering, to try to reduce the volume of formatted data to be loaded into the data warehouse of the service organization's BI system. To this end, a better understanding of the information structure and the flow of information among operational systems with the BI system in this service organization, as well as the social network analysis algorithm, will help us achieve this goal.

9.3.1 Information Structure of a Service Organization

In discussing the information structure of the large insurance service organization in question and the flow of information between the organization's BI system and its operational subsystems, we use the Unified Modeling Language (UML) diagram category. Application diagrams such as Use Case Diagram, Class Diagram, Activity Diagram, etc. are from the UML diagram category. After briefly examining these diagrams, we use one of the UML diagrams to draw the relationship between the BI system and the operational subsystems in an insurance service organization.

9.3.1.1 UML Model Diagrams

UML is a powerful programming language for showing and modeling system behavior, processes, and relationships between system components and their users in information systems. Consider an operational program at the executive level of a service organization. This program faces a huge volume of input data, output data, multiple interfaces, and very high-level components. In this program, displaying user interactions and communications, system processes, and system structure makes the system more observable. This further examination of the system by UML results in greater and better coordination between system components, which can increase the speed of program processing and save time and money.

This system modeling by UML and its business process modeling is the documentation of a set of steps and actions that are performed in a business process, and these actions are well modeled by its sub-diagrams.

Among the advantages of UML is its high flexibility, which allows us to model a system from two structural and behavioral aspects. And these two aspects are the main branches of the division of UML diagrams, namely structural diagrams and behavioral diagrams.

Structural Diagrams

Structure diagrams represent the static dimension of software systems and typically focus on the main structure of the system. These diagrams model the structure of a system well by showing operations, processes, identifiers, and relationships between system components. The real and observable relationships of the system are modeled by such diagrams, which are described below:

- Class diagram
- Package diagram
- Object diagram
- Component diagram
- Structure composition diagram
- Deployment and distribution diagram

Class Diagram:

A class diagram illustrates the object-oriented nature of a system. This diagram is considered the most common, most widely used, and most common diagram in UML, and its components include classes, dependencies, and collaborations as shown in Figure 9.1. Due to their static nature, these diagrams are mostly used for system construction and development. Classes in this diagram are represented by rectangles that contain information such as the class name, attributes, or properties of objects, and operators for how to perform work in these rectangles. The relationships between classes are shown as straight lines to indicate the dependencies between classes, with the numbers zero and one and an asterisk on either side of the line, where zero indicates no relationship, one indicates only one relationship, or an asterisk indicates multiple relationships. Or they are shown as a line with a diamond, which is called a part/whole relationship, with the diamond on the side of the rectangle of the whole.

Package Diagram:

A package diagram can be defined as a subset of a class diagram or, ultimately, as a technique of a class diagram. This diagram, as shown in Figure 9.2, shows the

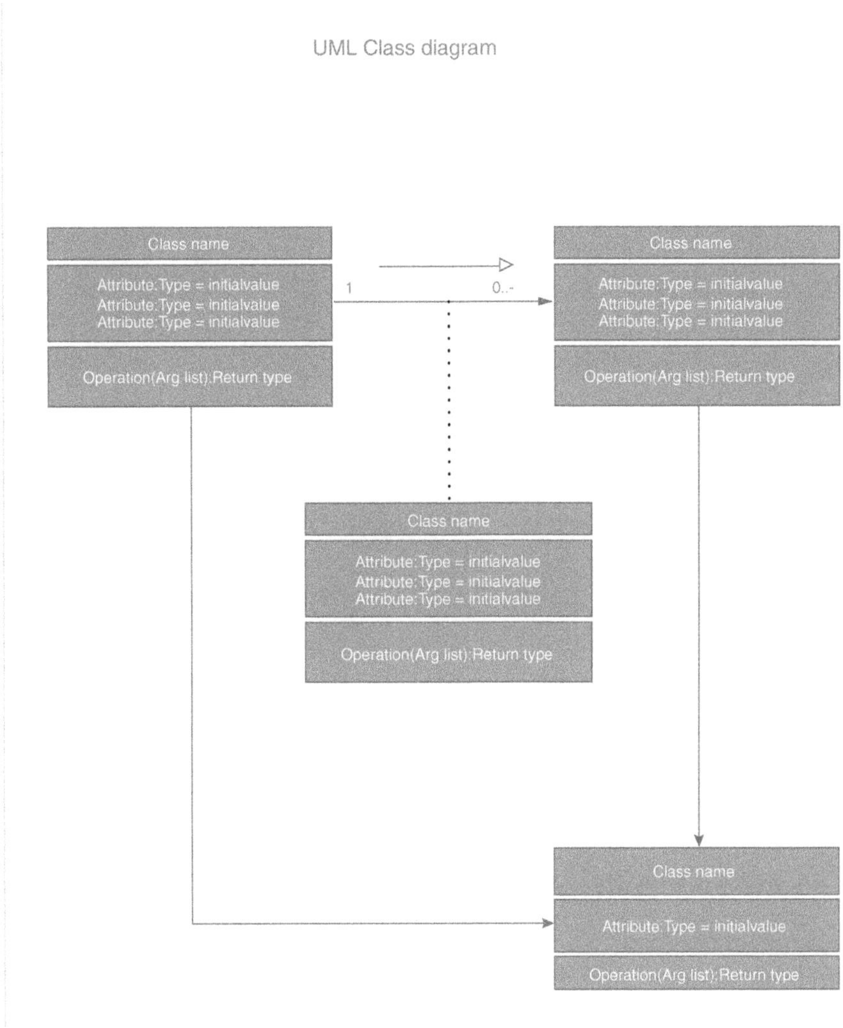

FIGURE 9.1 A class diagram.

subsystems of a system in the form of a package and shows its relationship with other systems, and in some way ignores the internal communications of the system.

Object Diagram:
This diagram involves time in the static structure of the system and can be used to test the accuracy of the class diagram.

Component Diagram:
This diagram models how the physical components of the system are organized. The command codes in the system databases are modeled in detail by this diagram, as illustrated in Figure 9.3.

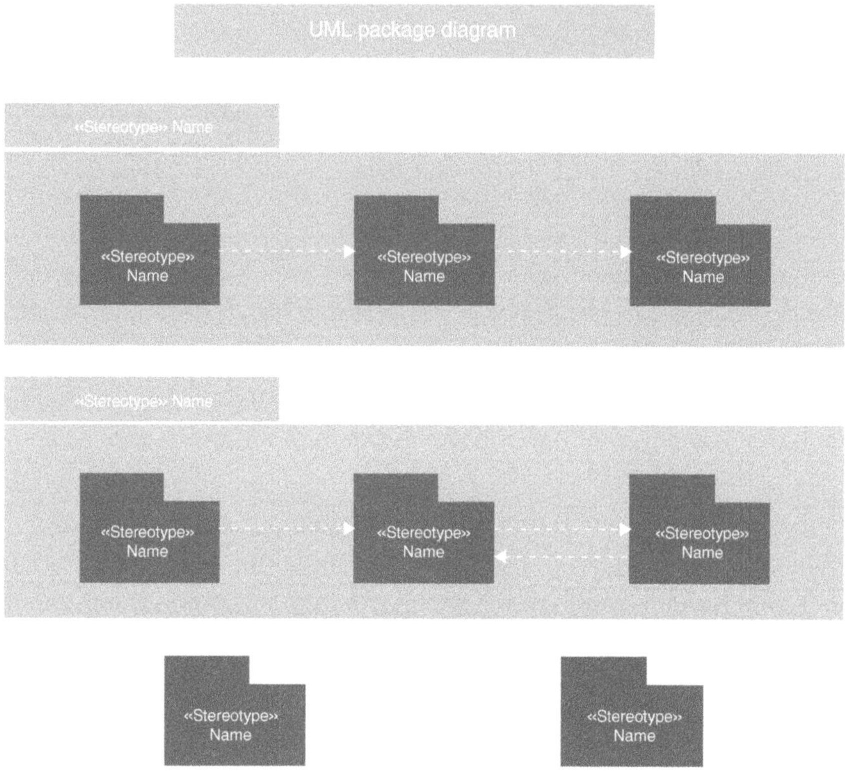

FIGURE 9.2 Package diagram.

A Composite Structure Diagram is used to show the internal components of a class.

Deployment Diagram:
This diagram can be used to describe the physical resources of a system. System resources are nodes, components, and connections (see Figure 9.4).

9.3.2 Behavioral Diagrams

In behavioral or interaction diagrams, the dynamic aspects of the system are modeled and displayed. Dynamic aspects of a system are the changing parts of the system. The following diagrams fall into this category:

- Use case diagram
- Activity diagram
- Sequence diagram
- State diagram
- Relationship diagram
- Interaction overview diagram
- Scheduling diagram

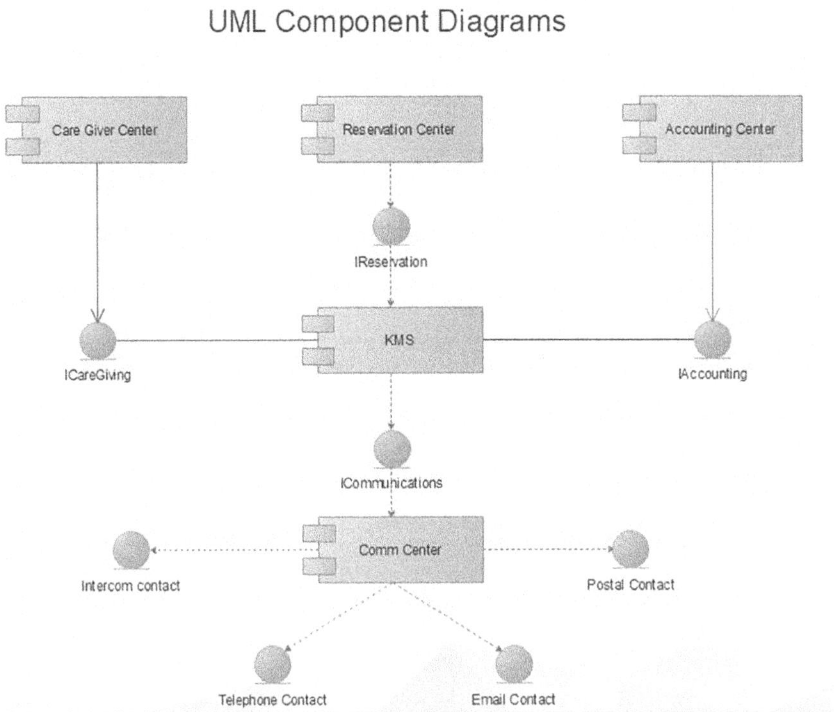

FIGURE 9.3 Component diagram.

Use Case Diagram:

Use case modeling is used to derive system objects and their interactions with each other. This category of diagrams is a procedure for analyzing a system in a way that addresses the nature of the business and how it works and has nothing to do with the way the system is built and implemented. This diagram has a view similar to the end user's view of the system and the system output, including behaviors, services, and responses, as shown in Figure 9.5. From the perspective of the system user or Actor, the use case diagram must produce valuable information, which requires the analyst to extract valuable metrics for the user. Use case diagrams contain agents, symbols used, and lines of communication. Agents are the external inputs of the system that play specific roles. For example, an employee or customer of a service organization can be called an agent that is located outside the system and is in contact with it. Of course, in addition to the aforementioned example of an agent, other systems that are in contact with the system modeled from the outside can also be considered as agents. Figure 9.5 shows the elements related to the system.

- Communicates Communication was established. It indicates the relationship of the agent with the use case. Naturally, this relationship should be beneficial for the agents.

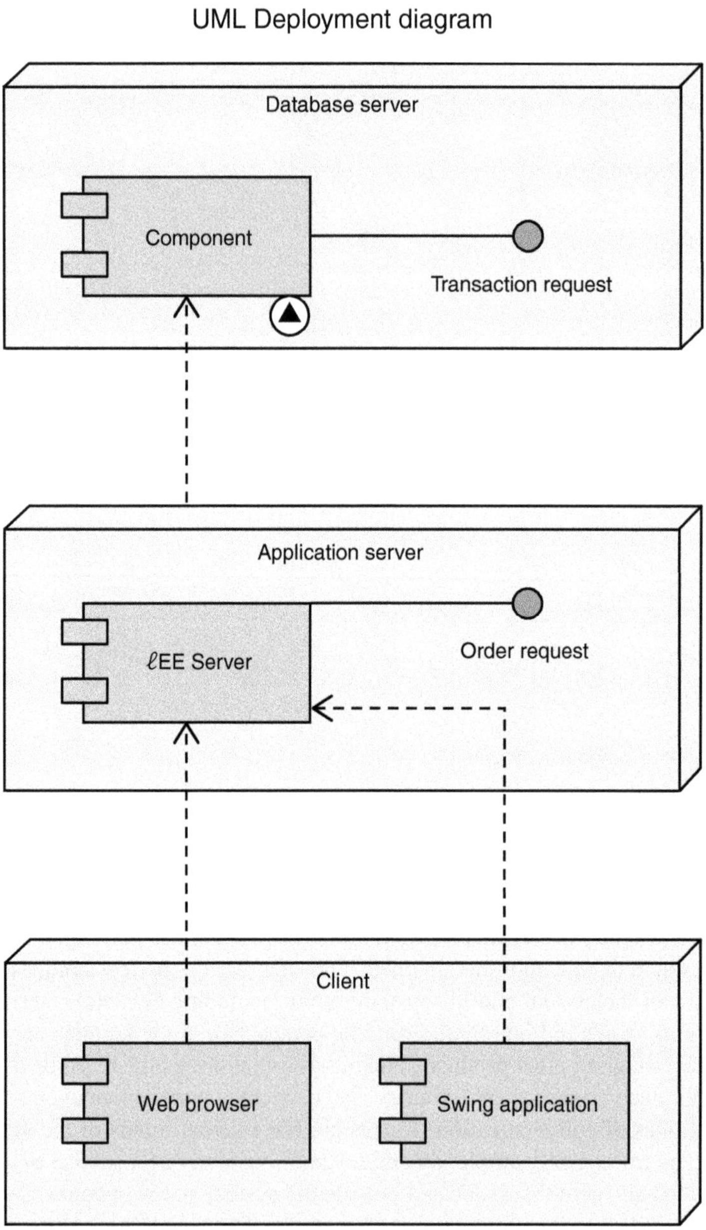

FIGURE 9.4 Component diagram.

- Includes Being included. Consider two systems that have similar behavior as use cases. This behavior is in the form of case behavior, not the entirety and totality of behavior.
- Extend Extend. This occurs when another agent affects the relationships of other agents in the use case when using the system. Annual insurance supplement tables that affect premium calculations are such relationships.

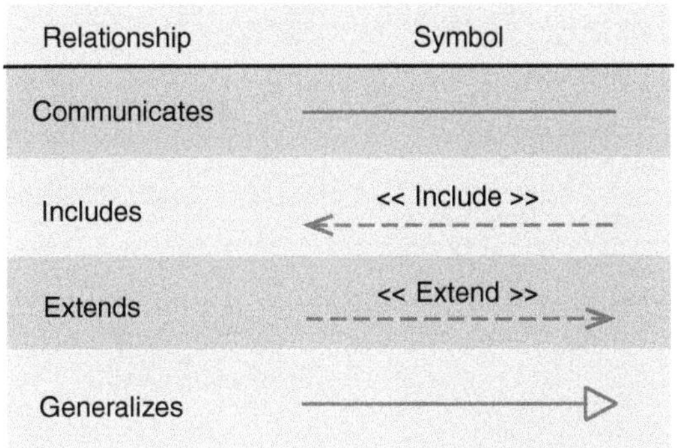

FIGURE 9.5 Use case diagram relationships.

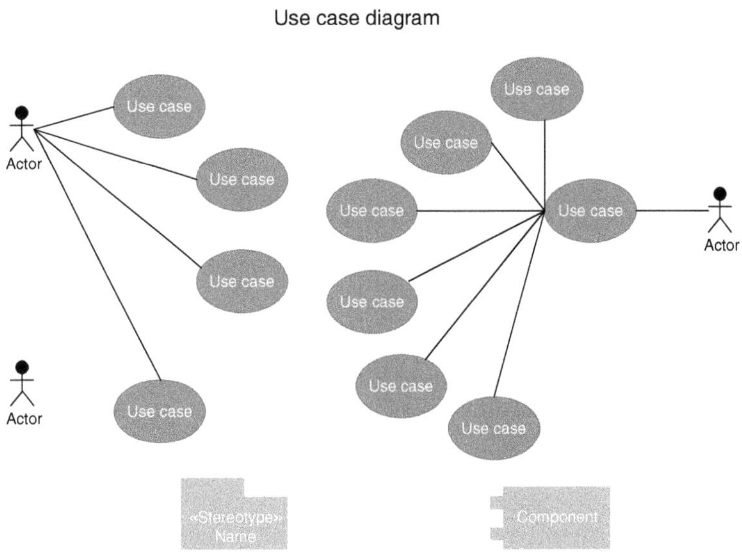

FIGURE 9.6 Use case diagram.

- Generalizes. The term generalizes the behavior of an agent by common agent criteria. In an insurance organization, it generalizes insurance marketers as well as insurance agents.

Figure 9.6 shows a use case diagram with the relationships described above and the agents using the system. Agents can have different roles, and several use case systems can be connected to each other.

Activity Diagram:
This diagram models the flow of control from one activity to another with four rectangular symbols with rounded corners as activities, arrows as events, diamond-shaped geometric shapes as diagram conditions, and finally thick lines that represent several

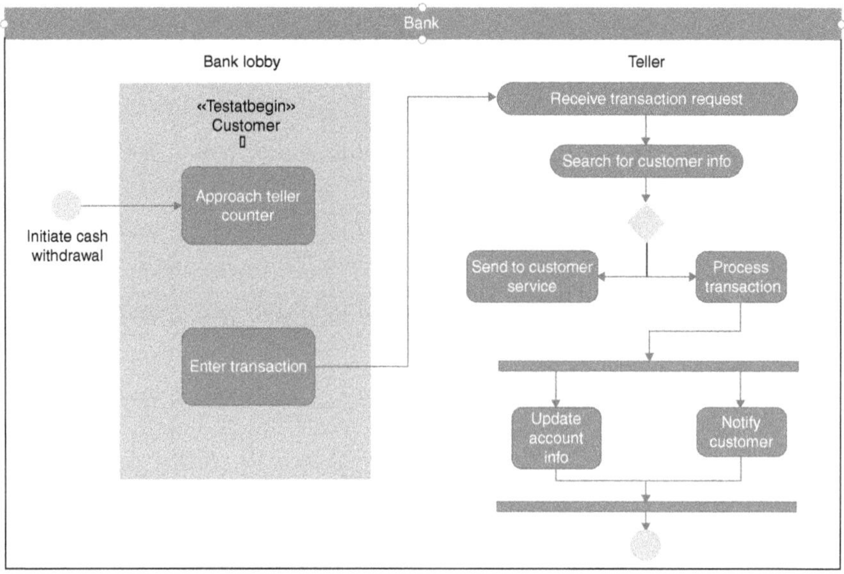

FIGURE 9.7 Activity diagram.

groups of events that proceed in parallel in relation to the system. An example of an activity diagram is shown in Figure 9.7.

Sequence Diagram:
As its name suggests, this diagram is used to display a sequence of calls and interactions between classes and objects in the form of message exchange to express the execution of system operations (see Figure 9.8).

State Diagram:
This diagram shows the behavior of the system in the face of an external stimulus. This diagram is used to model inactive objects in the system that are activated when an event occurs (see Figure 9.9).

Communication Diagram:
This diagram is used to model the interactions between different objects in a sequence. In other words, this diagram describes the static structure along with the dynamic behavior of the system.

Interaction Overview Diagram:
This diagram can be considered a combination of an activity and a sequence diagram in such a way that it models a sequence of activities.

Timing Diagram:
There are processes in the system that occur at specific time intervals. The timing diagram focuses on these processes and shows the passage of time from left to right rather than top to bottom in this system.

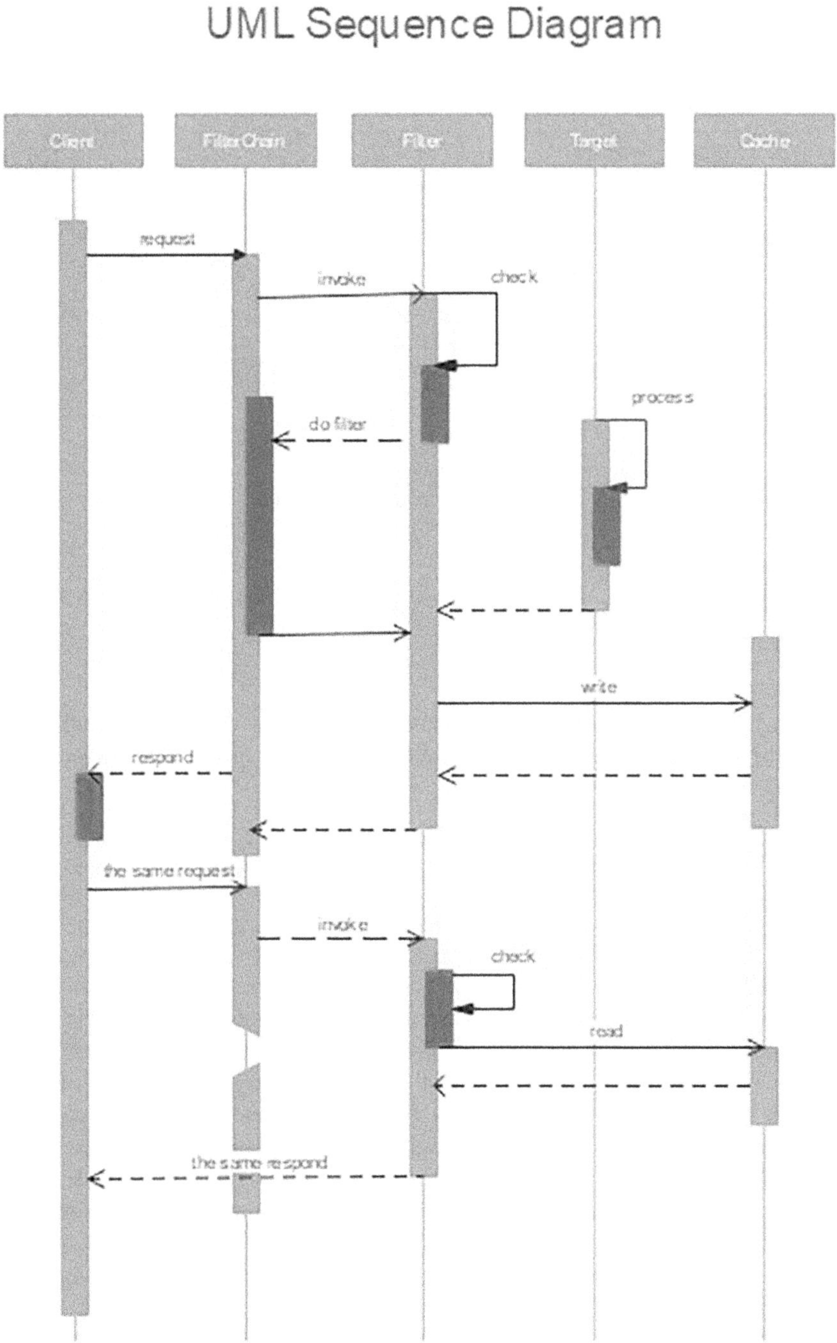

FIGURE 9.8 Sequence diagram.

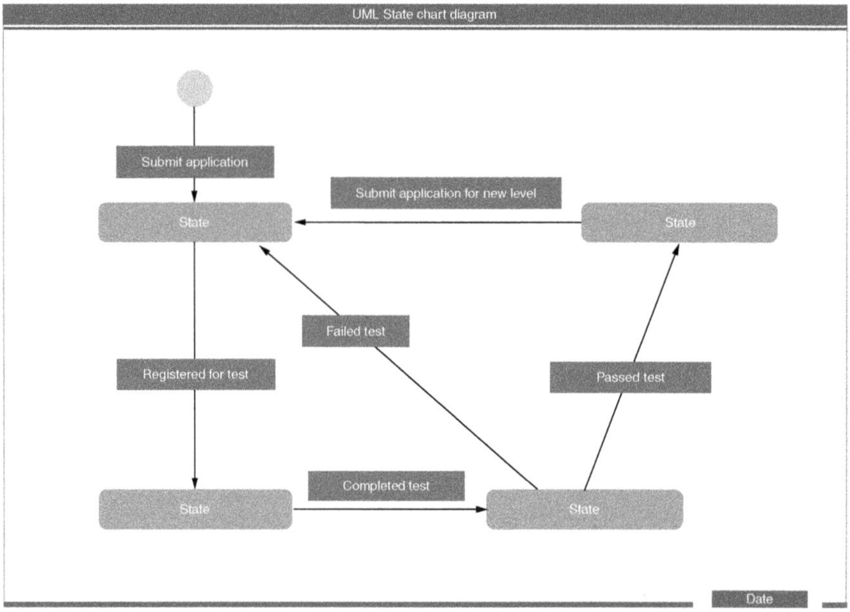

FIGURE 9.9 State diagram.

After this brief study of these UML information flow diagrams, we will try to model the information flow in an insurance service organization based on the use case diagram.

9.3.3 Modeling the Information Flow Structure in an Insurance Organization

To draw the information flow diagram in an insurance service organization, we consider the function of the BI system and the insurance operational subsystems. We model the BI system and the insurance operational subsystems as use cases in the diagram and draw the relationship between the components and objects of each of these systems with themselves and with each other. We model the role of senior managers using the BI system as a group of agents, the programmers of BI systems and insurance DSS subsystems as other agents, and the customers of the insurance system, insurance agencies, and branches as other agents in this diagram. Considering the breadth of this diagram based on the size of the insurance service organization, in this information flow, we associate three insurance DSS subsystems with the BI system of the insurance service organization and model two agents for each DSS: customer and agency, as indicated in Figure 9.10.

As we can see in Figure 9.10, in explaining the modeled diagram, it can be said: A senior manager in an insurance organization sends his request from the BI system to the BI manager. The BI manager provides his executive program to the BI programmer for reporting from the system. The BI programmer needs data from the

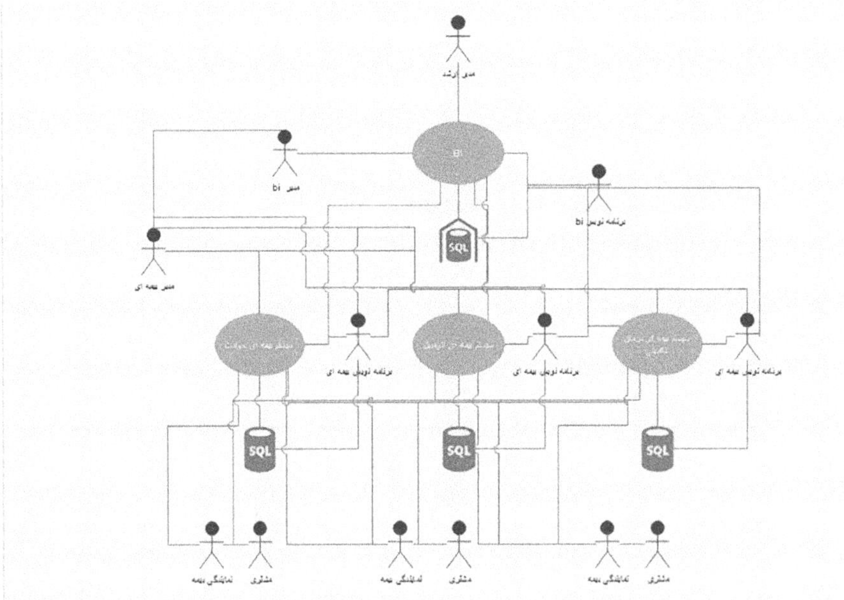

FIGURE 9.10 Information flow diagram of an insurance service organization.

databases of the supplementary medical insurance systems, the automobile insurance system, and the life and accident insurance system to structure the data and load it into the data warehouse. Since the BI programmer does not have direct access to the database of the organization's operational programs, the necessary coordination is carried out by the BI manager and the insurance manager to establish communication between the BI system expert and the expert of the organization's operational subsystems. After coordinating with the BI expert, the BI expert, with the help of the programmers of the supplementary medical, life and accident and automobile insurance systems, obtains the data required for ETL in the BI data warehouse from the operational subsystems database, structures it, loads it into the data warehouse, and provides the senior manager with the required output graph or chart through the BI system.

The data recorded in the operational subsystems database of this insurance service organization, including the supplementary medical insurance system, the automobile insurance system, and the life and accident insurance system, is obtained from the customer through the communication lines between the agency agents and the customer, which is shown in the figure, and is given to the system by the agency as one of the users of the operational insurance systems.

9.3.4 Social Network Analysis

As mentioned earlier, a social network is a collection of individuals or groups of individuals who are agents in the sense of agents, and there are some patterns of interaction, communication, or links between them. Friendships between real people, business relationships between companies, and marriages between families are all examples of networks that have been studied in the past. The following features and characteristics of these networks are essential for the study and analysis of these networks:

Agents as vertices and nodes in the network. The relationships between agents and the number of connections are called degrees. Clustering of sets in social networks and clustering coefficients according to the degree of correlations. Network analysis has a long history in sociology. Typically, network studies in sociology have been data-driven and involved empirical studies of real-world networks, often with the aim of determining the centrality in the network or the influence of various agents.

In these analyses, data-driven social networks were modeled and, given the nature of a social network, human studies were conducted on the characteristics of network actors as vertices of modeled nodes in graphs. Behavioral correlations between agents in these networks, human connections between agents in the form of binary connections, connections between two agents through a connection with a third agent without a direct relationship between the two agents, and boundary connections, so-called links between two graphs of a subset of a social network by agents or boundary vertices in these graphs, have been among the topics of interest to researchers in the sociology of social network analysis.

Recently, following the increased interest in the structure of social networks among mathematicians and physicists, numerous mathematically based models for social networks have been proposed. In these models, the agents and their connections with each other, the length of the connection between two agents, the average minimum distance between agents in clusters and subsets in a social network, the clustering coefficients in the subsets of the network, the degree and type of correlation of the members of a cluster, the uniformity or spread of data flows in clusters, etc. are all influential features in social networks that have found a numerical equivalent after being included in the presented mathematical models. For example, research has been conducted on the Internet and the World Wide Web in which a set of statistical features of networks and methods for modeling networks have been examined analytically or numerically. An important and fundamental result obtained from these studies is related to the number of relationships of agents with other agents, their so-called "degree". It has been found that in many networks, the distribution of degrees of agents is very complex, with a small number of agents having a large number of unusual relationships. Simulations and analytical work have shown that this standard deviation can affect how the network performs, including in predictions for how information is transmitted through the network and in the structure of networks to eliminate agents that have a small standard deviation in predicting their behavior in the network.

9.3.4.1 Mathematical Model of Social Network Analysis

After this overview of the literature on social networks and a brief discussion of the parameters and features that could be mathematically modeled, such as agents and their connections with each other, the length of the connection between two agents, the average minimum distance between agents in clusters and subsets in a social network, clustering coefficients in network subsets, the degree and type of correlation of members of a cluster, the uniformity or spread of data flows in clusters, we try to numerically model these features of social networks in a large insurance service organization as a large social network.

We denote the existing agents of social networks as N, where N can be N= {1,2,3,...,n}. N is known in any set as a node or vertex of the graph in that set. If our cluster is a database of insurance operational plans, the insurance profile of the insured individuals in it can be considered as an N factor, while if our cluster is a collection of individuals using a feature of this insurance service organization such as supplementary health insurance, then the insured individuals themselves, not their insurance profiles, are considered as N factors. In each of these sets, we represent the relationship between the agents with a degree M, meaning that the agents in the organization in question, which we modeled as a social network, or between the agents in the subsets of this social network as nodes, have a relationship of M={1,2,3,...,m}. For example, in this insurance service organization, the existing connections between agents in the main structure of the organization and the connections in the sub-groups of this organization can be expressed as follows: In the set of insurance agencies, the connections between branches and agencies can be an example of the use of M in the network analytical model. Direct connections between two agencies, star connections between multiple agencies, or connections between two agencies through a third agency are from this category. Or in the set of insurance system databases, the connection between the profiles of insured's can be an example of M, for example, in the complementary treatment database, all insured's of a production complex who use the complementary treatment service.

There is another category of connections in the structure of this organization, which we refer to as edge or link connections, and we denote them by $X_{(i,j)}$ where $i = \{1,2,3,...,I\}$ and $j = \{1,2,3,..., J\}$, and also $X_{(i,j)} = \{1,2,3,...,x_{(i,j)}\}$ represents the number of connections between subsets of this insurance organization. For example, the number of connections between the insurance system database set and the set of agencies and branches, and the BI system data warehouse is a type of this connection.

Another feature of this social network is the clustering coefficients in the subsets of the network and its relationship with the degree and type of correlation of the members of a cluster, and we denote it by C. The degree of clustering is usually measured using the average clustering coefficient.

We have $C(k) = K/N$, where K is the number of factors in a set with a relationship degree greater than the average of the set relationships, and N is the total number of set members. The closer this number is to one, the higher the coefficient of the model set and the higher the correlation and relationships of the set members, and vice versa. With the aim of improving the ETL process of data in the BI system data warehouse of this service organization, it can be claimed that for formatting the raw data of the insurance subsystem systems databases as subsets of this social network, if the clustering coefficient of these databases is high and close to one, we can use similar tables to format the raw data of these databases.

9.4 SABA INSURANCE SYSTEM

In this section, we use the Iranian Social Security Organization, considering its organizational dimensions and its desire to be a leader in information technology issues, which is also mentioned in the organization's mission and overall vision, as a platform for implementing our proposed method. First, we examine two insurance decision support systems of the Social Security Organization, namely Saba and the centralized one. By examining the data used in the decision support systems of the Social Security Organization, it is determined that the important and valuable data is the insurance profiles of insured individuals or insurance workshops, which are stored in raw and unverified form in the records of centralized databases and branches. Then, by showing two sample data sets from these systems, we first process them once in the normal mode in the BI system and measure the processing time, and in the second mode, we use the clustering capability in social network analysis to correlate and relate data, and thus increase the clustering coefficient in the databases of these two insurance systems, and after reprocessing, we measure the reprocessing time.

This operational system in the Social Security Organization is a desktop application based on NET that is used in the branches of the Social Security Organization. The processes related to registering insured information in various business branches, including income, inspection, and execution, are carried out in the branches. Inquiries related to the accuracy of information and claims of individuals in enjoying insurance benefits are checked in this program and registered in it after payment. The database used for the Saba system is on HP G8-G10 servers, which are named in the branches with DB (database) as the main database and DB2 (database 2) as the backup database. Social Security branches across the country are classified as Type 3, Type 2, Type 1, and Premium branches based on the number of insured and, accordingly, the workload and importance of the branch. Based on this division of branches, the equipment allocated to them and the volume of the database allocated to them are different. In this way, for type 2 and 3 branches, the storage space for the DB server is 200 GB, while this space for the DB2 server is 150 GB. In type 1 and premium branches, the storage space for the DB server is 400 GB and for the DB2 server, 300 GB. The insurance information of individuals, including the income, inspection, and execution sub-branches, which are still stored in the Saba system and in the branch database (the organization's goal is to move all platforms from the Saba system to the centralized system), is made available remotely and through telecommunication lines to the IT experts of the Social Security Organization, including the programmers of Saba or centralized insurance decision support systems and BI programmers to ETL the information in the BI system data warehouse. The security of this online connection is also established by hardware or software firewalls with Cisco or Juniper brands, and also ACS (Access Control Server).

Makes accounting-related matters more secure to control access to branch databases. In short, a connection is established between the headquarters and the branch, communication is established, access and access permissions are checked, and finally the branch database information is made available to the IT experts of the Social Security Organization headquarters for use.

The Saba application consists of the following three parts:

1. Data: Information stored in database tables according to the figures below. The structure of information tables in the Saba and centralized system databases is the same. This means that the structure of information tables is the same in Oracle and SQL Server databases.
2. Backend: The so-called source code of the program execution, which is displayed in Saba as follows. Saba programmers in the Social Security Organization work in this application layer and prepare the program for use by users and insurance experts in the branches of the Social Security Organization, with possible changes or modifications.
3. Frontend: This part is also called UI (user interface) and means the user interface that the end user of the program sees. The frontend of the Saba program is developed.

9.4.1 Centralized Insurance System

Due to the limitations and shortcomings that existed in the lack of centralization of the Saba program, for example, in the pension sub-branch of the Saba system, the possibility of establishing pension rights for an insured person in branches of two cities or two different provinces was not impossible. As a result, the organization's desire and willingness to centralize the services of the Social Security Organization prompted insurance programmers to write a new Java-based insurance application as a client-server called the centralized insurance system and make it available to users and experts of Social Security branches across the country. In the centralized decision support system, the database used to store insured information this time in centralized insurance sub-branches such as enrollment, pension, history, and medical documents is SAN (Storage Area Network) storage servers located at the organization's headquarters, whose current information storage capacity is 20 TB, which can be easily increased if the sub-branch, the so-called new sub-service, is increased. On this physical structure, namely the 20-TB SAN storage, Oracle database cluster3 software has been installed so that it can optimally respond to the high volume of information storage of millions of records of the Social Security Organization. The centralized insurance system application is installed on a server at the organization's headquarters, and in social security branches across the country. Branch users connect to the tools of this decision support system via the web, and actions related to data flow, such as data entry and obtaining output results from the system, are performed according to access permissions.

The centralized insurance system application of the Social Security Organization, like the Saba system, consists of three parts: data, backend, and frontend. In the first part, we have exactly the information tables with the structure for the Saba system, which we will avoid mentioning again. But the differences start from the second part. In the backend part of this application, as can be seen in Figure 9.11, these Java codes are written and developed by the centralized insurance system programmers at the organization's headquarters, and possible modifications and changes are made to it so that the end user of the application, who can be an insurance expert located in

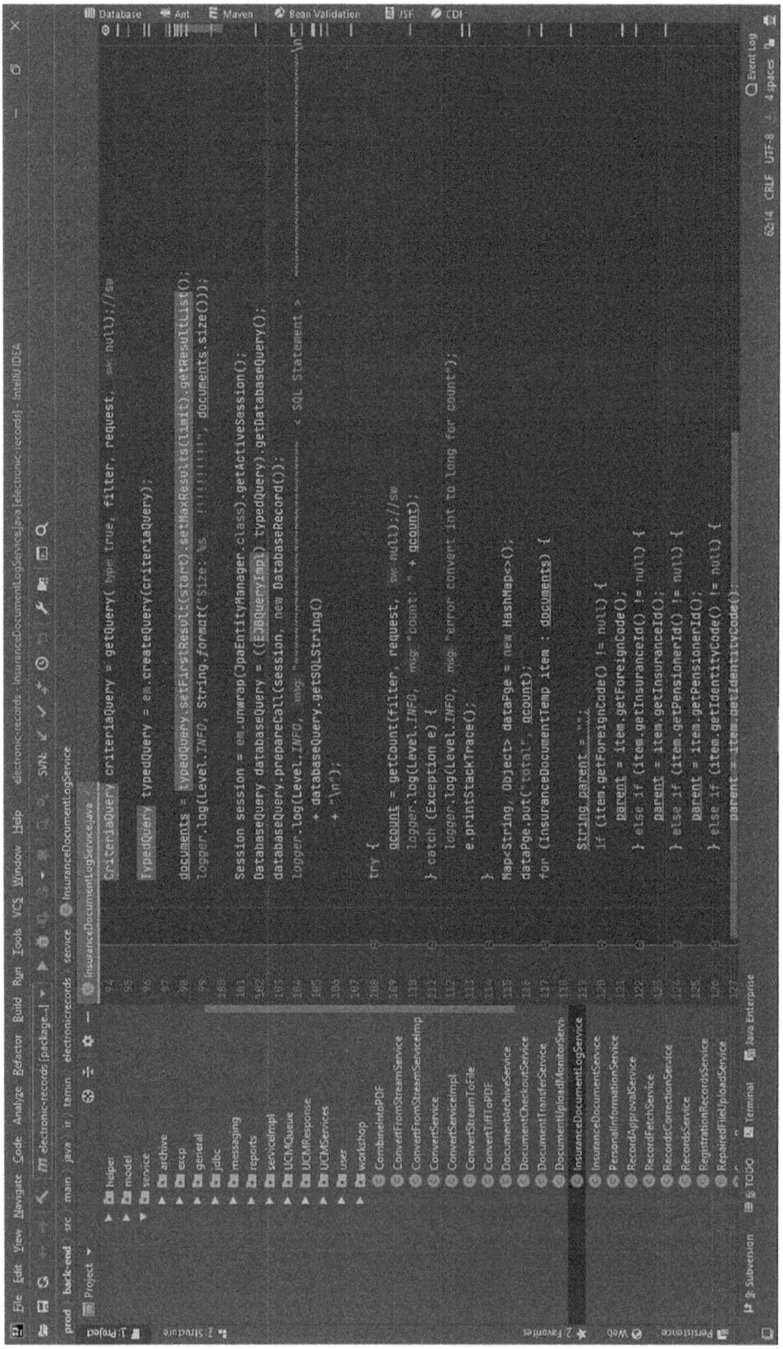

FIGURE 9.11 Java source code for the centralized insurance system.

branches across the country or an insured person of the Social Security Organization, can use the program.

In the third part of the centralized insurance application, namely its frontend, the end user accesses the centralized insurance system of the Social Security Organization using a graphical interface made available by the system programmer using JavaScript and has the ability to use various program tools in accordance with the access they have based on their role in the system.

The user is using the insurance Karstas system located in social security branches across the country. Actions such as calculating compensation payments for unemployment, pregnancy, etc. of insured persons or calculating insurance premiums such as purchasing insured military service years and adding their duration to the insured's insurance records are among the activities performed by branch experts on the centralized system. If the user is an insured user who connects to the centralized insurance system via the Internet and the web, they will see the graphical interface. As is evident in the menus in the figure, access to this system allows the insured to register many of their requests from the Social Security Organization.

9.4.2 BI System in the Social Security Organization

The BI system in the Social Security Organization has a similar application to the general application of the BI system in other organizations. Displaying macro-financial knowledge to senior managers of the Social Security Organization in the form of graphs and graphic diagrams is one of the applications of this system. The components of the BI system have been described in detail in the previous chapters. In this section, we will discuss the three parts of the BI program in the Social Security Organization, namely the input data to the BI, the ETL process of data in the data warehouse, and finally the output diagram. By giving a real example in the form of three figures, we will show the functioning of the BI system in the Social Security Organization.

Suppose that planning and decision-making regarding the community of social security physicians in the first half of 1400 in the city of Tehran, at the senior management level of the Social Security Organization, is our goal. Below are the steps for converting raw data from the decision support subsystems to formatted information in the data warehouse, and finally the BI output diagram for presentation to the senior managers of the Social Security Organization, as shown in Figures 9.12 and 9.14.

First, data related to doctors' visits and details of doctors' visits are obtained from the medical records service of the centralized insurance system. Two tables, t named tmdscore1400.doc_tr_taref and t1 named tmdscore1400.doc_tr_taref_details, are shown in Figures 9.12 and 9.13.

After obtaining the above data from the centralized insurance system database in the form of two information tables t and t1, in the ETL process of the social security BI system, the information of the two tables is mentioned and loaded into the BI system data warehouse in a third information table named PK. Figure 9.14 shows the ETL process and loading into the data warehouse.

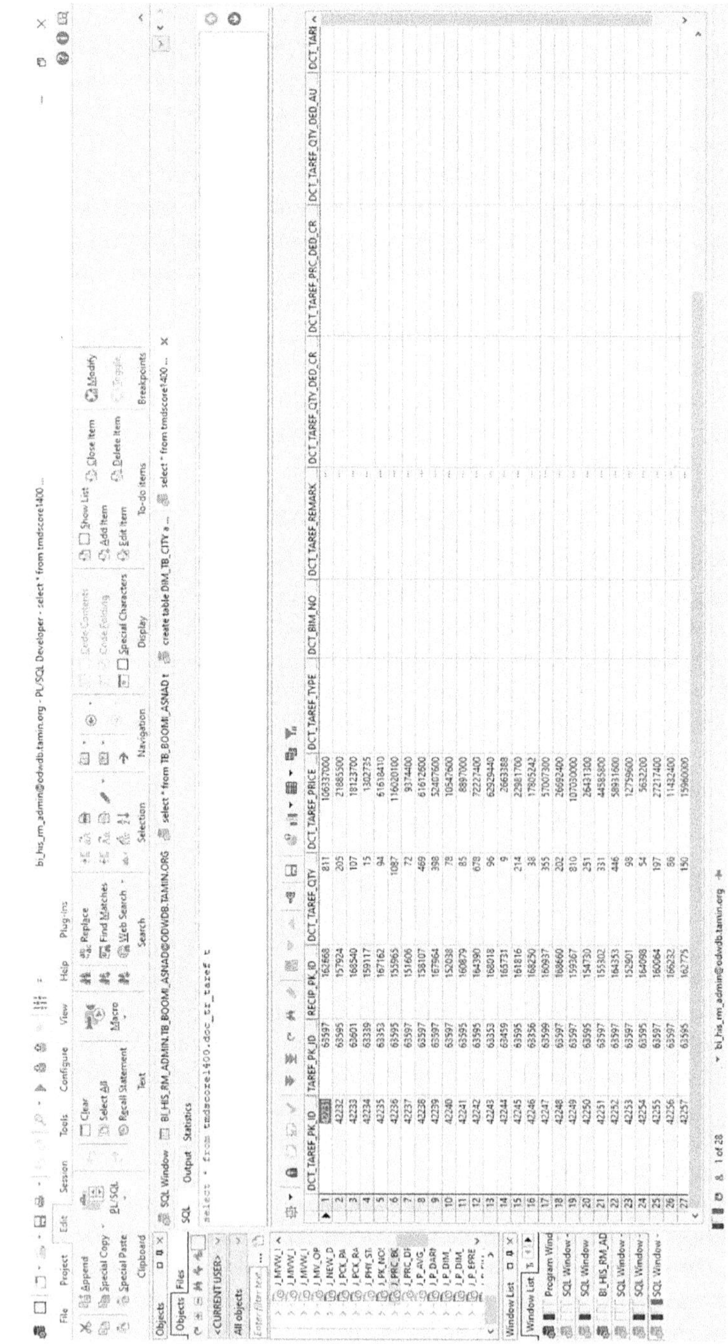

FIGURE 9.12 Physician visit data from the centralized insurance system.

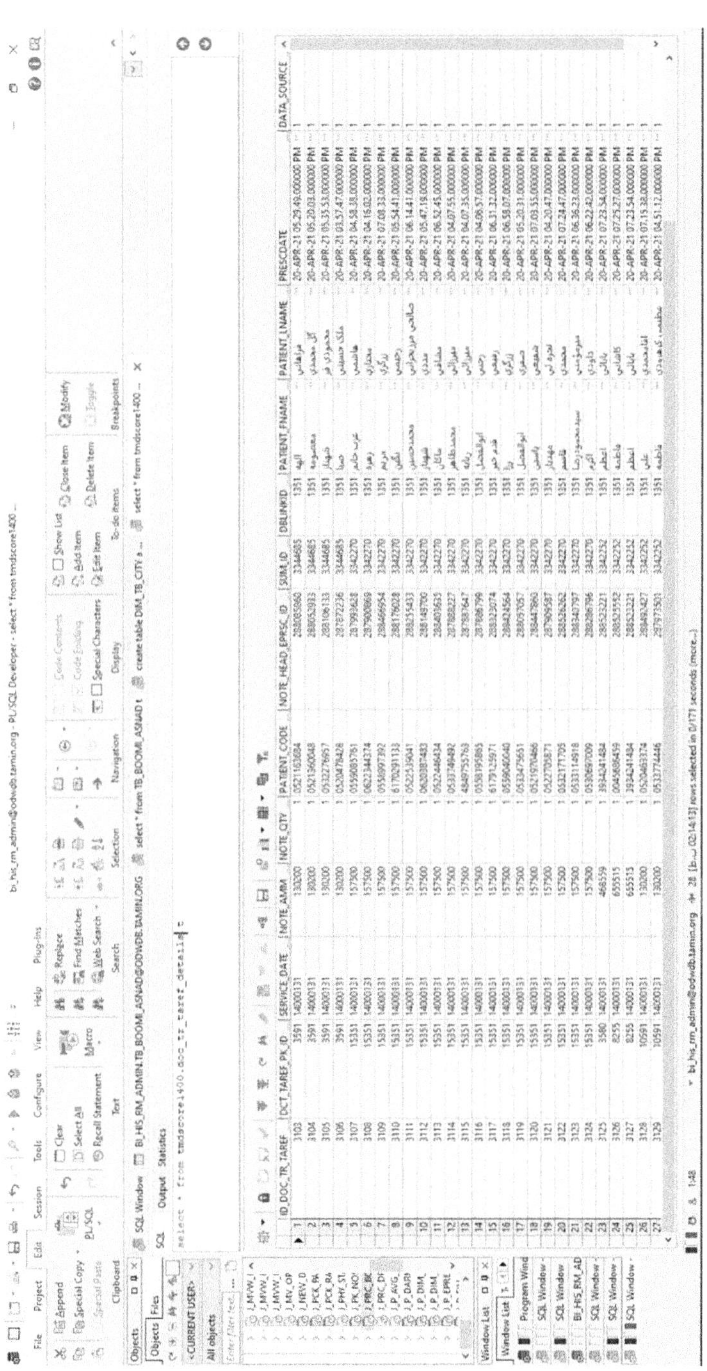

FIGURE 9.13 Physician visit details data from the centralized insurance system.

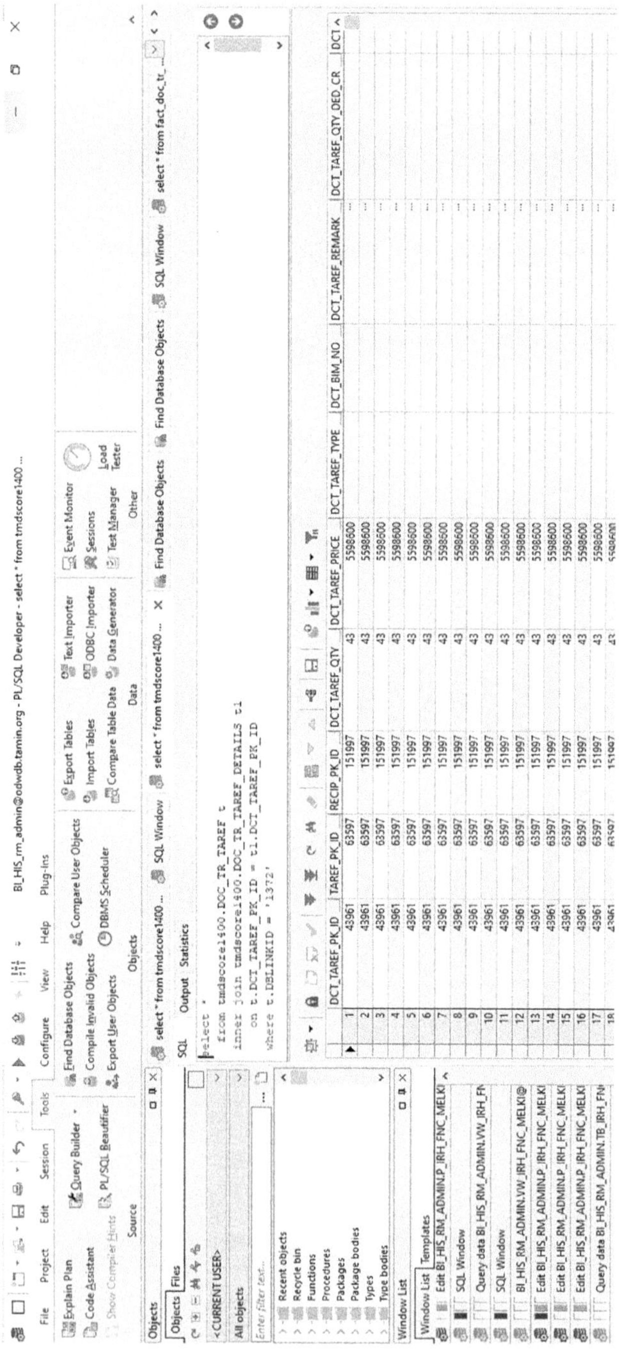

FIGURE 9.14 How to ETL in the BI system.

And finally, the required information output from the social security physicians community in the first half of the year 1400 is output from the BI system by month. These graphs contain the knowledge desired by senior managers of the organization for strategic planning and decision-making. For example, the tariffs paid to doctors in different fields.

9.4.3 Implementing the Proposed Model in the BI System of Social Security Organization

As we claimed in the previous chapter, by increasing data correlation and, consequently, increasing the connections between agents in a subset of a social network, we can increase the clustering coefficient of that set. Here, we consider the data stored in the sub-branches of the Social Security Organization's insurance decision support systems, such as the Saba or centralized system, as a set of social networks. With this analogy that we had about the similarity of these databases to social networks, we can improve the clustering coefficient of a data set by increasing the correlation between the data in a database, whether the fields of an information table or the connections between information tables. And then, in a real example, we examine the validity of our claim and show that increasing data correlation in a cluster increases the speed of ETL data from that cluster.

Here, we examine two examples of data tables from a centralized insurance system that we showed in Figures 9.15 and 9.16, namely, data related to doctor visits and doctor visit details as a subset of a social network. In these sets, data correlation can be increased by one of the appropriate indexing methods, logical partitioning of large tables, or database normalization and removal of duplicate data. Here, we test the increase in the clustering coefficient of these two clusters as a result of increasing data correlation by indexing these data tables.

In the real example mentioned in the last section, we measured the speed of program execution in the BI system during ETL of two data tables: doctor visits and doctor visit details in normal mode. The so-called cost of executing the ETL process in the pre-indexing mode and in the full index mode is equal to the numbers you see in the figure.

But in the second case, we index the two tables of doctor visits and doctor visit details, i.e., tables t and t1, before ETL. This means that the information entered in these databases in the first half of 1400 is seen in the program logic as 6 one-month periods before indexing and in full index mode, but with indexing on these two tables, the correlation of the data that is gradual and during these 6 months in these tables increases. In other words, it is seen in the program logic as a 6-month period. Naturally, due to the increase in the correlation and data connections in these data tables, which we modeled as subsets of a social network, the clustering coefficient of these clusters has increased. As can be seen in Figure 9.16, as a result of increasing the clustering coefficient of the two clusters t and t1, the cost calculated by the program during ETL of these two tables by the BI system has decreased. In other words, we say that the ETL speed has increased.

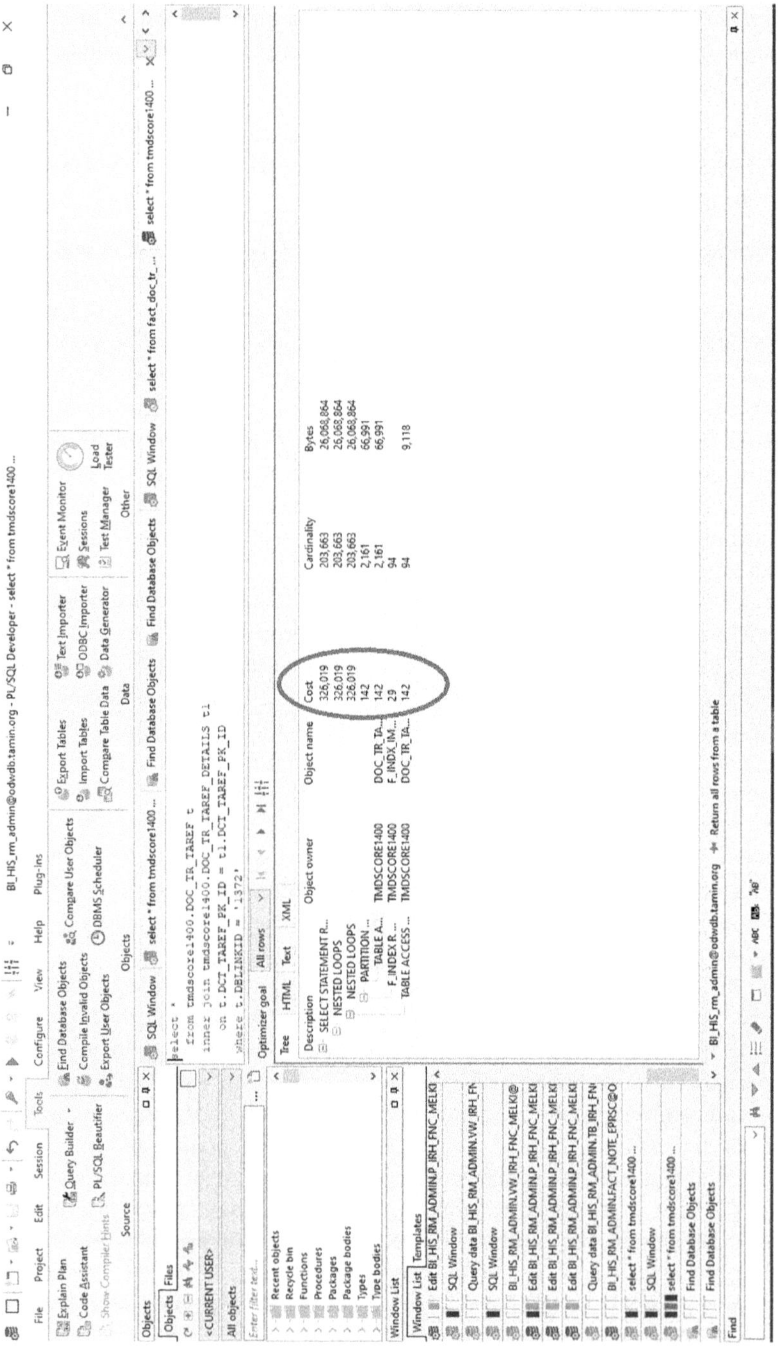

FIGURE 9.15 ETL speed in full index mode.

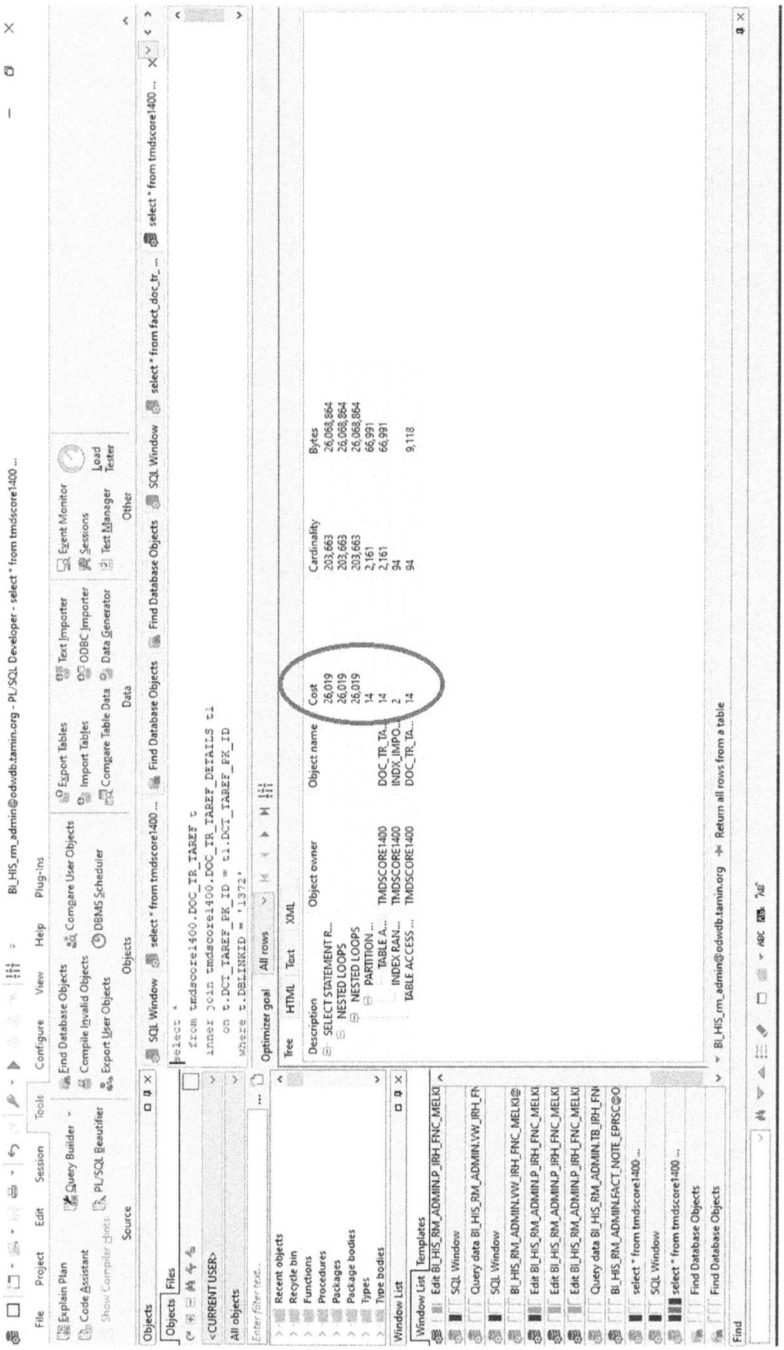

FIGURE 9.16 ETL speed after indexing.

By observing the cost numbers of the program in the red box and comparing them in Figures 9.15 and 9.16, the reduction in the time cost of executing the program is obvious. Based on the formula obtained in the third chapter, namely C(k) = K/N, as we remember, N was the total number of members of a subset of a social network, which in the example mentioned above was equal to 1,640,000 records. In the first calculation mode, i.e., the full index mode, we calculate the number K according to the definition of the formula, which was the number of records with a greater number of relationships than the average, as follows.

As mentioned in the program logic in full index mode, this large information table is seen as 6 one-month periods. Considering the nature of the records in these tables, it can be said that all members are related to each other. However, if we go into more detail, it can be seen that in full index mode, all members of a one-month period are related to each other, that is, about 273,000 records. According to the definition of the formula, the number of records that have relationships above the average makes the value of K one, otherwise the value of K tends to zero. The overall average of relationships in this example, considering the number of existing records, is equal to 820,000 relationships. As seen, the approximate number of relationships between members of this table in full index mode is about 273,000 relationships, which, according to the definition of K in the formula and the average of the relationships of the set of K in this case, tends to zero, and this information table is an uncorrelated table in full index mode, despite the completely similar data.

But in the indexing mode, the problem becomes completely different. In the logic of the program, after indexing, all 1640,000 records of the table are seen as a 6-month period, meaning that, due to the nature of the data being identical, all 1640,000 records of this set are related to each other. According to the formula K, it tends toward the number 1640,000, and with the numerator and denominator of the fraction becoming equal, the clustering coefficient tends toward one. In this case, by improving the speed of the ETL process, the situation of BI resource allocation in the Social Security Organization also improves. For example, about 5 TB less storage space in the BI program database.

9.5 CONCLUSION

In this thesis, we examined the general issue of the impact of information technology on the growth and development of large-sized organizations. We considered the Social Security Organization as a large organization as a platform to examine the role of information technology in it. Then, we divided the information technology branch into sub-branches such as business intelligence, BI systems, decision support systems, etc., and after providing definitions of them, we examined the role of each separately in the organization. We introduced UML diagrams to be able to show the data flow structure in the system, and we were able to show the information flow between insurance systems with a BI system and also the agents in a service insurance organization using the use case diagram. Then, we defined social networks and their algorithms, reviewed their history, and separated the capabilities of social network analysis that were most consistent with our goal in this thesis. By being more precise about the clustering capability in social network analysis and examining mathematical models, we obtained an applicable formula for this algorithm so that we could show the relationship between data correlation and clustering coefficient in the information technology systems of the Social Security Organization. We considered the Social Security Organization as a social network that included numerous graphs and subsets, and the decision support systems and the Social Security Organization's BI system are among these sets. And finally, by increasing the clustering coefficient in the databases of the Social Security Organization's information systems due to the increase in cluster correlation, we increased the speed of ETL data in the Social Security Organization's BI system.

In this research, we targeted the clustering capability in the social network analysis algorithm. Other features of social network analysis, such as sampling, diffusion analysis, centrality detection, community detection, and the impact of their modeling on information systems, can be the target of other research in this area. We examined the impact of the mathematical model of social network analysis on information systems, and we suggest that the impact of the sociological model of social network analysis on information systems should also be examined.

In the discussion of information systems and decision support systems or BI systems, we used the indexing method to increase data correlation between data tables. It is suggested that other methods mentioned, such as database normalization and logical partitioning of large data tables, should be examined to increase data correlation and its impact on increasing the clustering coefficient of data clusters and its result on data processing speed.

10 Loss Function Integrated with Internet of Things for Telecommunications Supply Network

10.1 INTRODUCTION

To establish physical communication between the telecommunications center and subscribers' homes and the process of providing a common tone, the system incurs normal operating costs, but due to technical and non-technical losses, additional costs are imposed on the system. Technical losses are usually caused by physical reasons, the quality of transmission lines, and the distribution system, but non-technical losses (NTLs) are caused by external actions of the system, the most important of which is the theft of copper cables due to their high $ value, and the other is the unintentional destruction of cable lines by individuals or other organizations. In this research, we want to use advanced Internet of Things (IoT) technology to provide an innovative solution to prevent this type of theft so that the damage is minimized.

In the process of establishing telecommunications connections between fixed-line subscribers and the relevant centers, the telecommunications company incurs high normal costs for physical and software setups. However, due to technical and non-technical losses, additional costs are also applied to the system, the most important of which is damage to the cable network. In this research, we are looking for a solution to assess the damage to minimize the damage to the cable network. For this purpose, the telecommunications network (physical communication routes between the telecommunications center equipment hall and cafes, posts, and homes of telephone subscribers). We consider the fixed telecommunications company (TSC) as a customer service delivery network and evaluate the service for maximum efficiency. In this network, nodes represent customer nodes, telecommunications centers are the center of the node, and edges are the connections between customers and centers. Any fault in each edge must be detected by the node and reported to the alarm center, so that the alarm center can inform the rapid response team about the presence of the fault and take corrective action as soon as possible. This mechanism is implemented under the Internet of Things system. IoT also offers huge potential when used to challenge the theft of copper cables. By placing devices embedded with sensors next to the trace cables, operators can detect the current in the wire. If the cable is hacked, the nearest device will cause a power loss at a specific point and

DOI: 10.1201/9781003740148-10

automatically alert the authorities of the location and time of the power loss. While the possibilities of this technology are exciting, using this type of solution in the network sector also presents various challenges that must be overcome before it can be adopted at a flexible scale. The non-invasive IoT solution will not interfere with any existing telecommunications infrastructure, instead using connectors that can sense through the cable without changing the existing functionality. This can overcome potential problems and have less downtime when using this solution. To evaluate this warning system, a loss function is used, which aims to minimize the mutual loss of the telecommunications system and customers due to the occurrence of a fault.

In order to establish physical communication between the telecommunications center and the subscriber's homes, a physical path must be established outside the telecommunications center. For this purpose, copper or optical cables are used, and these cables have different lengths depending on the distance from the telecommunications center to the desired location. Copper cables are of great interest to thieves due to their high value in $. And of course, unintentional destruction of cable lines by individuals or other organizations also occurs, which in addition to the high cost of providing and re-establishing cable lines, this disruption causes a huge loss of income for telecommunications for hours. Because when a cable is stolen and the phone is disconnected, the customer uses another tool to make a call, so we witness customer churn and therefore loss of profit. In this research, we want to minimize these losses by using the Internet of Things.

10.2 BACKGROUND

The occurrence of communication outages in telecommunications networks for any technical or non-technical reason causes heavy losses to subscribers of telephone line concessions as well as the companies providing these lines, and these outages occur in all countries for various reasons. Consequently, governments are always trying to find solutions to eliminate or minimize these losses. Among the programs that the telecommunications company has put on the agenda in most provinces of the country to eliminate this phenomenon and compensate for its damages, the following can be mentioned. This method is carried out on the entrance door of copper cable pools so that there is no possibility of cutting and stealing the cable. Also, a second door has been installed for the pool to reduce the possibility of opening those (Bagheri et al., 2018).

With the expansion of the fiber optic network and the failure of the copper cable network, the frequency of cable theft and disconnection of subscribers will definitely decrease. Therefore, replacing fiber optic cables with copper cables is one of the proposed solutions. However, the complete implementation of this plan requires a lot of time and is usually done for all the cable reversals that are being carried out. Usually, theft is done from inactive telecommunication cables, and theft of loaded cables is less common, which also does not take much time. Thieves only cut off a part of the copper cables and steal them, and in many cases, unfortunately, they even mistakenly cut optical fibers instead of copper cables, which only causes the disconnection of fiber subscribers, and the thieves will not gain anything.

Spacer literally means a spacer or distance holder, which is produced in various sizes and dimensions and plays a significant role in creating appropriate cable coverage. PVC pipe spacers are a suitable piece for maintaining the distance and precise arrangement between telecommunication pipes, and so on. This spacer consists of two parts, a base and a top, which create a regular and integrated gap between telecommunication ducts. This product is mainly used for urban telecommunication installations and operations and industrial projects such as refineries and petrochemicals. Over time, cables corrode due to contact and proximity to chemicals and environmental conditions such as humidity, air, substances such as sulfates, etc. This phenomenon will have a very destructive and harmful effect on the cable network. However, if a spacer is used, a coating is created on the cables that will protect them from weather conditions. This component is usually used in the construction of aerial and terrestrial networks to significantly increase work efficiency. These components are very useful tools that, when used, prevent the cable from moving in the network and, in other words, keep it fixed.

In order to facilitate access of specialized workers to the transmission network channels and to continuously inspect the wiring and make necessary changes and repairs, engineers install specific points in different parts of an underground cable route. In such cases, electricity and telecommunications manholes are usually used, which are actually the places where the workers can access the transmission network. A manhole is actually a cylindrical or rectangular chamber made of concrete, polyethylene, or brick, and their construction details are carefully designed and constructed according to the different applications it can have in different systems. The cover of an electricity and telecommunications manhole can also be made of cast iron, steel, composite, or polyethylene. Electricity and telecommunications manholes can be used in large projects. For example, in extensive power transmission networks or complex optical fiber transmission networks, the power and telecommunications manhole can accelerate the construction and completion process of the project, fully support and protect the wiring, and ensure the longevity and security of the project in the long term (Fazlollahtabar and Ahmadiahangar, 2024).

In coordination with the local law enforcement unit and the security of the Tehran Province Telecommunications Company, integrated patrol activity can be increased, and preemptive measures can be taken to prevent the theft of urban telephone cables. In the meantime, the role of security institutions such as the police becomes more prominent, because the damage to telecommunications cables is outside the telecommunications circle and is noticed by such institutions. Because there is a pond at a distance of 200 to 250 meters. Therefore, in a city with a large number of telecommunications centers, there are many ponds on a street. Sometimes the thief cuts off the sub-cables. Most of the people who are arrested by the police or telecommunications security are addicts. Some places don't even steal. They don't know, and since they don't have a place to sleep and the weather is cold, they go into these ponds and light a fire, which catches fire there and destroys the cable.

As mentioned earlier, if thieves succeed in stealing cables, they need suitable places to collect and store the cables so that they can melt the wire coating and extract the copper inside. By periodically inspecting waste storage facilities and conducting legal and judicial investigations in case of suspicious cases, preventing the purchase and sale of metal waste without obtaining the necessary permits, and increasing supervision by the Department of Industry, Mines and Trade over metal smelting workshops, access to these facilities can be minimized, thus minimizing the risk of cable network theft and consequent damages to the telecommunications company.

Emphasizing that telecommunications is one of the organizations that create the communication platform and a wide national and international network, if telecommunications equipment is exposed to threats and theft, by cutting off communication in villages, the outskirts of cities, and roads, many problems and damages are imposed on society, and the feeling of security is minimized. As a result, it is necessary to prevent such damages with the wisdom and cooperation of telecommunications personnel and interaction with the police force. Therefore, in order to prevent and reduce the destruction and theft of telecommunication equipment, expert meetings should be held, and the problems and reasons for the thefts should be examined. Some of the thefts that have occurred have required knowledge of the efficiency and price of the facilities. Therefore, in order to counter the possibility that some profit-seeking individuals may steal telecommunication equipment due to administrative problems with contractors, efforts have been made to apply maximum care and caution in hiring personnel over the past year, and extending contracts with contractors who have had administrative and personnel problems has also been avoided. Any transportation of wires, cables, and other electrical equipment by the company's contractors must be carried out with a valid exit slip. It was requested that vehicles carrying this equipment be controlled by the police and that appropriate planning be made to control the city's metal smelting workshops to prevent the melting of stolen cables (Fazlollahtabar et al., 2013).

Telecommunications contractors are required to replace the cable within 72 hours in the event of a breakdown or failure, and ensuring the cables are guaranteed for this replacement and minimizing the time of telephone outages.

Researchers analyzed a queuing system with a general scheduling function for controlling telecommunications network fault detection and compared the performance of different service scheduling policies for telecommunications network detection based on the probability of loss and average waiting time with numerical examples, and finally, they applied the results obtained to control the telecommunications network route. Thus, by controlling the time priority, they provided the exact delay time of the demand estimate and proposed methods for the steady state probability of the queue length and the waiting time distribution coefficients to control the priority of the loss. A queue is a place where service-receiving units wait for their turn, and queuing theory is related to systems in which the arrival of customers and their service are random. By locating the best location for the server, the mobile server moves to the location of the requester point in the given network, and after completing the service at that location, it always returns to its original location before servicing other requests. Queuing theory is one of the most important applications of probability theory and stochastic processes. Considering the social and economic importance of queuing in the fields of engineering, telecommunications, transportation, and computer systems, and social systems, many mathematicians have been researching this field for years, and now, due to its increasing development and expansion, extensive research is being conducted in queuing theory.

Researchers examined the current use of two governance tools – privatization and regulation – and proposed that subsidies and public information be promoted as a solution to eliminate electricity losses, considering the existing political obstacles. Turkey's reforms in electricity distribution through government deregulation have focused on providing high-quality services while reducing customer costs. Increasing electricity efficiency is considered the most important way to reduce electricity prices and improve electricity quality. Since non-technical losses are recognized as the main barrier to achieving these goals, the Turkish government has taken many steps to address both issues. The study identifies two governance tools (economic regulation, privatization) that have already been implemented and two additional tools (subsidies and public information) that should be used to address the inefficiency issue. Economic regulation is defined as the economic activities of controlling public and private entities in order to positively change their behavior in the market and ensure the provision of essential services to consumers without destroying the ability of the market to function. Privatization refers to the sale of public assets to the private sector. Subsidies refer to the provision of public funds from the government to local authorities, public and non-governmental organizations, or individuals to achieve specific policy outcomes. Public information is a policy tool that involves providing information to a target audience about a specific issue and policy to change their opinions and behavior. In the following sections, each tool is applied to the problem of theft, and their effectiveness and political acceptability are discussed.

Economic regulations typically involve a set of standards and expectations aligned with political objectives. For example, they may require a registration process to obtain a license that allows an activity in the market. In addition, the regulatory process involves oversight that can result in the revocation of a previously issued license if the required standards are not properly maintained. Current regulatory policy is very effective in achieving and even exceeding the goals and reducing the rate of NTL theft.

Governments can use advertising and indirect organizations to provide goods and services that were previously provided directly by the government. This process is known as privatization. Privatization is also used to support the private sector through public policies, such as public information aimed at changing the behavior of the public sector. Typically, deregulation and targeted tax deductions or credits are examples of this form of privatization.

Grants are defined as payments made by a national government to sub-governments, non-governmental organizations, units, or individuals to achieve specific policy outcomes. Instead of directly providing goods or services, governments shift the responsibility for the performance of a task to third parties. The tool of grant-making can be used effectively to support research efforts that provide information to help facilitate decision-making processes, which has the potential to promote more effective policy.

Public information tools are defined as a method of informing target audiences about a specific policy issue by increasing knowledge and expanding belief systems in order to promote positive behavioral patterns. The main goal of this approach is to achieve public recognition that theft-related activities are a criminal offense that carries specific criminal penalties. There are different methods of public information.

Predicting customer behavior in the telecommunications industry is one of the important tasks in relationship management. The competitive business environment forces companies to build a common prediction model for individuals to increase their profits.

Researchers examined the cause and effect relationship between telecommunications infrastructure, economic growth, and development in selected African countries and also estimated the triple effects of telecommunications infrastructure, economic growth, and development in the region. This analysis considers a panel of 46 African countries from 2000 to 2015. The empirical results suggest a long-term bidirectional relationship between telecommunications infrastructure, economic growth, and development. Communication infrastructure drives economic growth and development in Africa and vice versa. Therefore, comprehensive and holistic policies should be promoted that promote digitalization, economic growth, and development simultaneously in Africa. Increasing telecommunications infrastructure will encourage overall production and living standards to move in the same direction in Africa.

The researcher recommends that first, the communication sector, economic growth, and the need for development should be simultaneously promoted because there is a mutual causality between them to promote simultaneous sustainable development in Africa. Second, developing and promoting policies in the telecommunications sector that improve fixed-line deployment and Internet penetration, as access is still low in Africa, will boost economic performance. Further development and expansion of fixed-line networks will have a broad impact on other electronic communication tools (mobile and fixed-line broadband Internet services) as the interconnection of communication services still relies on efficient fixed-line networks. Third, empirical results showed that the Communication Infrastructure Index (CIT) is of high importance for economic performance, growth, and development, and vice versa in Africa; therefore, African countries should adopt more policies that strengthen the efficient performance of communication services and promote inclusive economic growth and development.

Governments in emerging economies, usually when domestic capital is reluctant to invest in technologies that have the potential to disrupt the existing ecosystem, must intervene and leverage the power of these investors. The role of governments in promoting innovation in technology can be mentioned as providing the initial conditions, organizing industrial networks and social capital, and empowering institutional intermediaries. However, empirical results are mixed, with some believing that the long-term orientation of the government will lead to innovation, while others question its ability to foster innovation. Networked public-private partnerships constitute independent but interconnected factors, and the government, as an important factor, can act as a facilitator and coordinator of the formation of a network of public-private partnerships. In particular, it has been found that the government can contribute to the success of the public-private network by coordinating cooperation and creating a coherent network or by playing a brokering role in facilitating knowledge exchange and recombination. Most importantly, previous research has shown that the specific structure of the network can cause creative disruption in entrepreneurial activities.

The case study shows that (1) the government-initiated network has a more centralized structure at its beginning. (2) As the network evolves from a government-initiated process to a government-coordinated one, increasing cohesion and structural fragmentation facilitate knowledge production. (3) As the network evolves into an embedded structure, the government and GSI usually maintain their positions during the evolution.

The principles and conceptual concepts for building a new class of automated systems, namely online decision-making systems for communication networks, are presented. Their distinctive feature from other systems is that part of the management decisions for communication networks can be realized automatically. The basis of the proposed class of online decision-making systems is hierarchical fuzzy positional networks, fuzzy mathematical programming methods, and hierarchical fuzzy inference methods. This approach allows one to effectively change rational management decisions in external operational conditions. The proposed architecture of the online decision-making system is focused on the implementation of several agents, the main elements of which are intelligent agents and managers. Intelligent agent architecture allows one to implement online decision-making systems in communication networks using specific cases, software, design, support systems, and online decision-making systems. The evaluation of the effectiveness of the proposed approach, compared to traditional probabilistic methods, showed significant benefits and demonstrated the ability to use the proposed class-Flancing-Theory-Supervisory-System in real-world performance.

Customer retention is one of the most important challenges in the communications industry. Companies may find it essential to predict customer churn for the success of their operations, as a thorough investigation of churn can be a vital tool for customer retention. In addition, the push-pull framework is used to examine the impact of features on customer behavior and personnel behavior. Also, behaviors or illegalities are analyzed. The results show that logistic regression has the highest prediction accuracy, and the percentage pressure coefficient of churn is one of the most important factors affecting the customer. Communication service companies use a wide range of tactics, including creating a service portfolio (bundling), offering a variety of programs to meet customer needs, and offering discounts to ensure customer loyalty and attract new customers, thereby increasing market share. Ironically, such tactics increase competition among service companies, and therefore, it is crucial to understand customer behavior and preferences, provide appropriate customer service, retain customers, and minimize churn and flexibility, to the detriment of customers who switch from one provider to another within a given period of time. In fact, customer retention by minimizing customer churn is one of the most important issues for telecommunications service providers.

With increasing customer demand, the market is now mature and at the same time increasingly competitive in the telecommunications industry. In addition to tactics, companies also use a wide range of strategies. One of these strategies is Customer Relationship Management (CRM), which works in the field of long-term profitable development. Relationships with key stakeholders and customers, the development of recent digital systems, information technologies, and the most advanced machine learning and statistical technologies provide ample opportunities for telecommunications companies to have a better understanding of their customers, which is very important for the development of an effective CRM system. Today, intense competition and pressure to reduce costs have caused many companies to focus on improving their CRM.

The recent development of information technology has led to the accumulation of a large amount of data in many companies. Such data, if managed correctly, has significant potential to predict challenges and ultimately strengthen operations and better and more informed decision-making. To do this, data must be properly cleaned, organized, balanced, and analyzed using appropriate tools (Mehrmanesh et al., 2013).

Despite the rapid development of the telecommunications industry as one of the leading and innovative industries, the acceptance of more and more users of telecommunications and Internet services, and, in general, increasing dependence on the commercial and social environment of these people, these services, which depend on the supply chain management of telecommunications companies, are rarely considered. While the high pace of innovation and the importance of providing reliable services in this industry pose many challenges for the sustainability of the country, there is almost no comprehensive study on the supply chain. Due to the various defects that occur in the fixed telephone telecommunications network and the resulting communication disruption, it is very important to study this issue, evaluate the amount of economic loss caused by it, and also provide a way to adjust it. A telephone outage or failure of a part of the network is always one of the most fatal economic issues in telecommunications companies. Cable theft and outages caused by storms and floods, especially in Mazandaran province, are one of the main challenges in providing services to subscribers. Therefore, this research aims to design an alarm structure based on the Internet of Things and an evaluation method based on the loss function for analyzing defects in the fixed telephone telecommunications supply network.

10.3 RESEARCH METHOD

The emergence of new Internet of Things technology has a profound impact on people's lives and work methods. The terrestrial Internet of Things cannot achieve global coverage. However, satellite communications, due to their wide coverage, can be integrated into the global coverage and thus greatly expand the business scope of the terrestrial Internet of Things. Satellite networks are a powerful complement to the terrestrial network and an important part of the global communication system. Compared with Earth-orbiting satellites and medium-orbiting satellites, low-orbiting satellites have many advantages, such as low latency, low transmission loss, and integrated global coverage. Therefore, the combination of low-orbiting satellite networks and terrestrial networks can effectively expand and supplement the shortcomings of the terrestrial communication network and solve the problem of global coverage. The initial ideas of the satellite Internet of Things were based on traditional satellite communications. Based on the traditional satellite system, various applications are added, such as integrating the IoT perception layer for data collection, and long-term tracking and monitoring of the IoT device delivery area.

10.3.1 Problem Definition

In recent years, routing algorithms and application strategies for low-orbit satellite networks have been studied by many researchers. According to the topological structure, they can be divided into static routing and dynamic routing. According to the different data transmission modes, the strategies based on wired and wireless connections are divided. And also, depending on the location of the calculated route, they are divided into centralized routing and distributed routing. The goal of this system is to simplify the time-varying topology in order to simplify the routing algorithm. Therefore, in this research, we use the narrowband Internet of Things.

Narrow Band Internet of Things (NB_IOT) and Long Range Radio (Lo Ra), which are widely presented, are used to analyze the relevant system characteristics and network architecture in the Internet of Things. Pointing to the characteristics of dynamic network topology and unstable connections, the related factors on the performance of the routing protocol are analyzed. The rapid development of the Internet of Things has created higher requirements for terminal connections. Traditional terrestrial cellular networks and short-range communication technologies (such as Bluetooth, Wi-Fi, etc.) cannot meet the needs of the Internet of Things for long-distance communication.

Therefore, the low-power network LPWAN (Low-Power Wide-Area Network) with low bandwidth, high power, long distance, and wide connectivity came into being. Low-power wide-area networks, abbreviated as LPWAN, are a class of wireless communication technologies that enable low-data-rate connections over a wide geographical area with low-power (battery) consumption. The aforementioned features have made these networks a suitable option for connecting sensors and making cities and industries smarter.

Lpwan is evolved from Long-Term Evolution (LTE), simplifying and optimizing it. For this reason, it is very suitable for real-time data transmission application scenarios in the Internet of Things. The NB-IoT network architecture is basically the same as the 4G network architecture, but it has been simplified for NB-IoT, such as omitting the physical connection control channel, reducing the transmission bandwidth, simplifying the transmission process, etc. At the same time, some functions are also improved.

10.3.2 Modeling

NB-IoT is an LPWAN technical standard defined by the 3rd Generation Partnership Project (3GPP) standardization organization. It uses licensed frequency band technology, which greatly reduces the power consumption of traditional terrestrial mobile networks and solves the problems of long-distance IoT transmission. NB-IoT is not a new protocol and follows the basis of 4G network architecture. The characteristics of IoT determine that the terminal must have low-power consumption capability. It connects directly from node to node through the air interface, supports control plane and user plane optimization processes, and has the function of link speed control. Serving Gateway (SGW) supports the random access type of NB-IoT and is very practical and convenient for the routing of data packets or data transmission to the next layer. The server collects the IoT data received from various access networks, analyzes, organizes, stores, and uses the data to provide various processed data to customers. Lower circuit height, lower transmission loss, lower transmission delay, and higher transmission power are the advantages of the NB_IOT architecture, and we introduce the characteristics of the NB-IoT system from three aspects of narrowband communication technology, optimized network transmission, and low-power terminal design.

Inside the cable boxes, the cables are unloaded on terminals installed in the boxes and sorted toward different posts. From each box, the cables enter several posts in different neighborhoods and alleys with a smaller number of pairs. From each post, the wire pair goes to the common house or shop with the number of connections available in the post (10 or 20 connections). According to the explanations given in the second section, at any point of the existing physical path between the telecommunications center and the beep delivery location, a connection can be disconnected. In this research, we want to consider the telecommunications network as a network with nodes and edges and, using IoT technology, inform the telecommunications center and technical support of any problems in the nodes or disconnections of the edges in the shortest possible time. For this, we assume all subscribers as nodes and consider the telecommunications center as the center of the nodes and assume the cable path connecting the two to be an edge. An overview of the proposed network is shown in Figure 10.1.

Now that the structure of the network under study has been designed, we are looking to apply a method to calculate the damage to the system during a disturbance in the telecommunication network. In this study, a loss function based on clustering is used. Clustering analysis plays an important role in many engineering disciplines such as pattern recognition, system modeling, image processing, data mining, etc. Clustering algorithms try to divide a data set into several subsets (groups or clusters) such that data in the same cluster is more similar to data in other clusters. So far, many clustering algorithms have been introduced for different applications, which are classified into two groups: probabilistic/fuzzy and stable clustering (Shahri et al., 2013).

10.3.3 Clustering Based on Loss Function

In order to develop various clustering algorithms, Kohnen proposed Learning Vector Quantization (LVQ) fuzzy clustering neural network. Cellular clustering neural network (FCNN) is based on LVQ, which is one of the machine learning algorithms and is classified in the family of artificial neural network models and neural computation. Due to its simplicity and efficiency, it has attracted more and more attention. As a generalization and completion of LVQ, FCNN network clustering was proposed. The elements of this model are introduced below:

$$X = \{x1, x2, \ldots, xN\}\text{: data}$$

$$xi = [xi1, \ldots, xid]T\text{: input sample set}$$

$$m\text{: number of clusters}$$

$$V = [v1, v2, \ldots, vm]\text{: cluster center}$$

$$vi = [vi1, \ldots, vid]T\text{: the center of the i-th cluster}$$

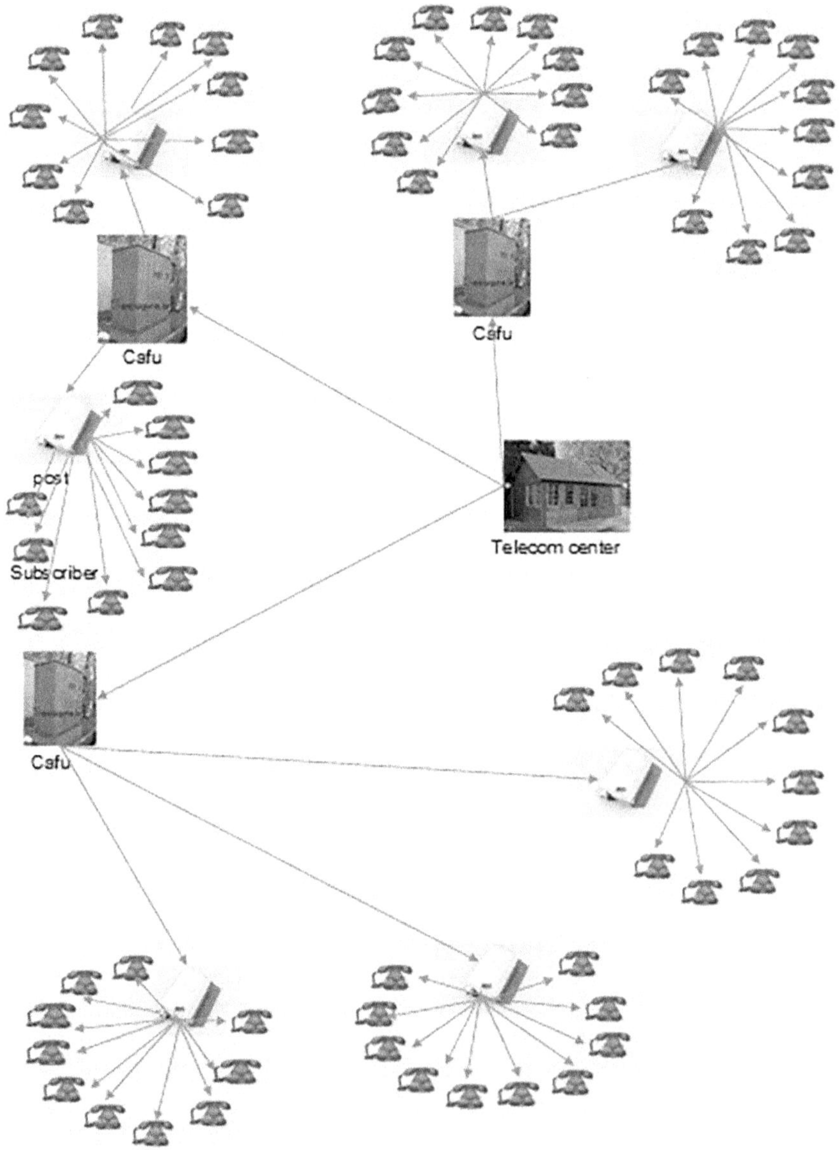

FIGURE 10.1 Telecommunication network.

In fact, FCNN, as one of the objective functions, is to minimize the following function:

$$J_m = \frac{1}{2}\sum_{i=1}^{M}\sum_{j=1}^{N}\frac{1}{\sigma^2}C_{ij}U_{ij} \| X_{ij} - V_i \|^2 \tag{10.1}$$

$$i = 1, \ldots, M, \quad j = 1, \ \ldots, N$$

Using the following update rule:

$$V_i(t+1) = V_i(t) + \eta \sum_{j=1}^{N} \frac{1}{\sigma^2} C_{ij}(t) U_{ij}(t) \left(X_j - V_i(t) \right) \quad (10.2)$$

where η is the learning rate, which is usually set to 1

$$C_{ij}(t) = \exp\left(-\frac{d_{ij}^2(t)}{2\sigma^2} \right) \quad (10.3)$$

$$d_{ij}(t) = \| X_j - V_i \| \quad (10.4)$$

$$U_{ij}(t) = \frac{C_{ij}(t)}{\sum_{h=1}^{M} C_{hi}(t)} \quad (10.5)$$

Uij(t) is the membership of the jth sample in the ith cluster.

In practical applications, real data is often corrupted by distant nodes. For example, one of the main disadvantages of FCNN is its sensitivity to distant nodes. The clustering performance is seriously degraded by its sensitivity to distant nodes. Since there are often values of distant nodes in real data, this disadvantage limits its wide application. Therefore, how to develop RFCNN for its practical applications is very important. To increase its robustness, FCNN should consider the distance robustness measure instead of the conventional L2 distance measure. As a robust loss function, we choose the Vapnik- μ loss function as the robust loss function in this study.

Objective function performance based on the loss function:
The Vapnik electronic sensitivity loss function is used because of its simplicity and effectiveness. Here, the clustering neural network uses the electronic sensitivity reduction function to measure the distance. With the help of the Vapnik electronic sensitivity loss function, we redefine the objective function in FCNN and accordingly present the new robust fuzzy clustering neural network, RFCNN. Our experimental results show its strong robustness against distant nodes. In particular, when the Vapnik- μ loss function is established, determining the parameter ε becomes a fundamental problem. Our RFCNN automatically adjusts this parameter for the sample data to avoid this problem. The distance is used for its simplicity and relatively low computational burden. However, when the real data contains more distant values, this criterion leads to the FCNN being overly sensitive to distant nodes. In order to make the FCNN robust to distant nodes, we need to choose robust measures in the objective function. In our robust RFCNN fuzzy clustering neural network, the sensitivity-ε-panic loss function is used as the distance measure.

The sensitivity- μ -vapnik loss function is defined as follows:

$$|f|_\varepsilon = \begin{cases} 0 & |f| \le \varepsilon \\ |f| - \varepsilon & |f| > \varepsilon \end{cases} \quad (10.6)$$

where μ is the insensitive parameter. When μ=0, the sensitivity- μ -Vapnik loss function becomes the absolute error loss function.

The new objective function in RFCNN is defined as follows.

$$J_{m\varepsilon} = \frac{1}{2}\sum_{i=1}^{M}\sum_{j=1}^{N}\frac{1}{\sigma^2}C_{ij}U_{ij}\left\|X_j - V_i\right\|_{\varepsilon_i} + \gamma\sum_{i=1}^{M}\left\|\varepsilon_i\right\|^m \tag{10.7}$$

where

$$\left.\begin{array}{l} \left\|X_j - V_i\right\|_{\varepsilon_i} = \sum_{k=1}^{d}\left|x_{jk} - v_{ik}\right|_{\varepsilon_{ik}} \\ \left\|\varepsilon_i\right\| = \sum_{k=1}^{d}\left|\varepsilon_i\right| \qquad \gamma \geq 0 \end{array}\right\} \tag{10.8}$$

$\varepsilon = \left(\varepsilon_{i1}, \cdots \varepsilon_{id}\right)\left(\varepsilon_{ih} \geq 1 \cdots d\right)$ denotes the vector corresponding to the insensitive parameters in the ith cluster, and m (> 0) denotes the exponential index described earlier and is taken to be 1 for simplicity.

This new objective function consists of two parts: the first part is the use of the electronic insensitive loss function, Vapnik's, which is used to remove distant nodes in real data.

The second part is the penalty for insensitive parameters. The introduction of this penalty period gives the new objective function the following advantages:

A given objective function should minimize the first term in (10.7) to its minimum value and at the same time minimize the second term in (10.7). The new objective function (10.7) reflects this feature well, and the parameter γ is used to adjust the balance between the first and second terms.

10.4 DATA ANALYSIS

We conduct our research for the telephone subscribers of the University of Babol's telecommunications center and examine the telecommunications company's revenue in relation to the number of active telephone subscribers. The variability in the number of subscribers is due to disconnections due to various reasons, including cable theft.

These surveys were conducted on four days, on 08/01/2019, 08/12/2019, 08/14/2019, and 08/17/2019, and on each of these days, due to the disconnection of the subscribers' telephone lines, their performance and, consequently, the company's revenue decreased.

On 08/17/99 with 33,567 subscribers, the total revenue was $1018,906,315
On 08/16/99 with 33,522 subscribers, the total revenue was $997,354,531
On 08/14/99 with 33,517 subscribers, the total revenue was $852,852,753
And on 08/13/99 with 27,285 subscribers, the total revenue was $544,495,126.

With a rough calculation, it can be seen that by disconnecting the phones of 6282 subscribers for 4 days (from the 13th to the 17th), the revenue decreased by $474411189, which is a significant figure. The experimental studies are aimed at comparing the clustering and resilience performance of RFCNN and FCNN. In this regard, we studied experiments.

The data consists of three separable sample groups, and each group consists of 20 samples. In this experiment, σ is set to 1, and the same initial cluster centers (1.75, -0.25), (1.45, 1.55), and (4.5, 0.5) are used in FCNN and RFCNN, respectively.

In order to better understand the changes of the above clustering results, in the case of nodes, we use $\Delta V = \| V2 - V1 \|_2$ to measure these changes, where V1 represents the actual cluster center matrix, and V2 represents the corresponding cluster center matrix. It is clear that the smaller ΔV is, the stronger the corresponding clustering capability algorithm is.

Results in Table 10.1 clearly show that the clustering performance of FCNN suffers severely in cases where the data contains distant nodes. With the increase in the number of nodes, ΔV in FCNN becomes larger, while the clustering performance of RFCNN does not change drastically in the presence of nodes. When $\gamma \geq 4$, ΔV in RFCNN is very small. Even with the increase in the number of outliers, the accuracy of the data is maintained within ΔV. This table also shows that when $\gamma = 4$, RFCNN shows the best clustering performance, and when $\gamma \geq 8$, ΔV remains unchanged, and the cluster centers are also acceptable. However, when $\gamma \leq 2$, with the increase in the number of outliers, the obtained cluster centers become unacceptable, although ΔV still changes slightly.

The dataset used here is the dataset of the cutoff on different days. In this experiment, σ is set to 1.5, and the same initial cluster centers (3, 0.4) and (5, 0.3) are used in FCNN and RFCNN, respectively, and the maximum number of iterations is 20 and 300 in RFCNN and FCNN, respectively. By adding different numbers of nodes at the coordinates (6, -1), we can robustly investigate the capabilities of FCNN and RFCNN for the nodes. The ΔV changes of FCNN and RFCNN are listed in Table 10.2.

TABLE 10.1
Performance Comparison for FCNN and RFCNN

				ΔV			
	RFCNN				FCNN		
Node	**$\gamma = 0.01$**	**$\gamma = 0.5$**	**$\gamma = 2$**	**$\gamma = 4$**	**$\gamma = 6$**	**$\gamma = 8$**	**$\gamma = 100$**
0	2.7413	1.236	0.8759	0.1008	0.1364	0.2886	0.1349
3	2.9041	1.236	0.8784	0.1008	0.1364	0.2886	0.1349
6	2.9041	1.236	0.8784	0.1008	0.1364	0.2886	0.1415
9	2.9041	1.236	0.8784	0.072	0.1364	0.2886	0.9821
12	2.9041	1.6786	0.8784	0.1251	0.1364	0.2886	0.9826
15	2.9041	1.6786	0.8784	0.1251	0.2074	0.2886	0.9828
18	2.9041	1.6786	0.8784	0.1251	0.2074	0.2886	0.983

TABLE 10.2
Performance Comparison between FCNN and RFCNN on the IRIS Dataset

	ΔV								
	RFCNN					FCNN			
Nodes	$\gamma = 0.001$	$\gamma = 1$	$\gamma = 4$	$\gamma = 6$	$\gamma = 8$	$\gamma = 10$	$\gamma = 15$	$\gamma = 20$	$\gamma = 100$
0	2.8897	2.8287	2.183	0.0874	0.0523	0.0764	0.0839	0.0951	0.2598
2	2.8649	2.8287	2.2286	0.0874	0.0523	0.0764	0.0839	0.1422	0.2598
4	2.8649	2.9619	2.2744	0.0874	0.0523	0.1128	0.0839	0.0951	0.2453
6	2.8649	2.9619	2.3013	0.1206	0.0681	0.1128	0.0839	0.0951	0.2377
8	2.8649	2.8229	2.3478	0.1206	0.0681	0.1128	0.0839	0.0951	0.2381
10	2.8649	2.8229	2.3482	0.1206	0.0681	0.1128	0.0839	0.0951	0.2541
12	2.8649	2.8229	2.3482	0.1626	0.0681	0.1128	0.0839	0.0951	0.2705
14	2.8649	2.8229	2.3482	0.1626	0.0681	0.1128	0.0839	0.0951	0.2874
16	2.8649	2.8229	2.2079	0.1626	0.0681	0.1128	0.0628	0.0951	0.3048
18	2.8649	2.8229	2.6054	0.1467	0.1074	0.1128	0.0628	0.0951	0.3227
20	2.8649	2.8229	2.6054	0.1467	0.1074	0.1128	0.0628	0.0951	0.3411

Table 10.2 shows that the clustering performance of FCNN is affected by the number of nodes in the data. As the number of nodes increases, ΔV in FCNN becomes larger, while the clustering performance of RFCNN does not change significantly in the presence of nodes. When $\gamma \geq 6$, ΔV in RFCNN is very small and almost unchanged over a wide range of node numbers. This table also shows that $\gamma \geq 8$, with different γ, ΔV is almost invariant, and the cluster centers are also relatively reasonable. However, when the number of nodes increases, the changes in ΔV are still small, and the obtained cluster centers become unacceptable. From the above two experiments, it can be seen that when γ is very large, RFCNN almost obtains a more reasonable clustering result than FCNN. Although ΔV of FCNN may be less than RFCNN when the nodes are very small (see Experiment 4-1), as the number of nodes increases, the superiority of RFCNN over FCNN is clear.

10.5 DISCUSSIONS

We consider different points in the telecommunication network, which are the homes of subscribers, and with the help of the clustering algorithm, we classify these points geographically. We denote the points in the network where the outage occurred by x and the center by v. The clustering algorithm takes the centers as data and divides them into three groups by considering the average distance between the points. This algorithm aims to reduce the loss by giving a series of criteria and grouping them in such a way that the centers are placed together so that the least loss is incurred by the system. By putting the geographical coordinates of the outage set in the function, the amount of loss is shown according to the amount of income. We also saw that the FCNN clustering algorithm has a series of disadvantages and shortcomings for remote points. Therefore, in order to effectively overcome the disadvantages of

FCNN sensitivity for distant points, we have reconstructed its objective function by introducing the sensitivity reduction function and Panic K, and accordingly, we presented the robust fuzzy clustering neural network RFCNN and its clustering algorithm. Our experimental results show that the clustering results of RFCNN are more suitable than FCNN clustering.

10.6 CONCLUSION AND SUGGESTIONS

In this study, we investigated the telecommunication cable network and evaluated the telecommunication supply network serving IoT-based subscribers using the loss function. In order to effectively overcome the sensitivity loss of FCNN for remote nodes, by introducing the Vapnik sensitive loss function, we reconstructed its objective function, and based on this, the RFCNN neural network clustering and its clustering algorithm performance were derived. Our experimental results showed that RFCNN can obtain reasonable clustering results, and its robustness is also very high compared to FCNN. The following suggestions are recommended to future researchers for the development of the model and its applications:

- Implementation of the method designed in this study for power grids to prevent regional outages
- Development of a Bayesian clustering method to investigate interacting states in communication uncertainty
- Development of a loss function method with pricing for telecommunication services with a certainty compensation approach

11 Reinforcement Learning for Smart Energy Management

11.1 INTRODUCTION

In recent years, due to the increasing complexity and developments of the global community, the energy factor has played a fundamental role in the economy and politics of every nation, and careful monitoring and outlook of the energy sector and adopting an appropriate strategy are the main pillars of maintaining the stability and political power of every country. Energy consumption is the most important factor in air pollution and climate change. For this reason, the optimal use of energy in the economic development process has always been considered an important goal in sustainable development. This is possible by optimizing and improving the methods of exploiting resources and processes of energy processing, conversion, and transfer, as well as implementing energy management.

Achieving energy management goals will entail obtaining new equipment or technologies and using the desired level of energy throughout the organization, which will ultimately minimize energy costs, waste without affecting quality and production, and minimize environmental impacts. On the other hand, in order to protect non-renewable energy resources and their optimal use, which also belong to future generations, society should be diligent in preserving them. Especially in the industrial sector, by establishing an energy management and auditing unit and implementing measures such as standardization, training programs, and financial mechanisms for energy management, as well as incentive programs, an effective step can be taken in the rational and optimal use of resources and energy with high efficiency.

Energy is the ability and capability of humans or other objects to do work; this work is called power. From a scientific perspective, energy can be defined as: Energy is the amount of energy consumed in a given time for a task. Since the proper and appropriate use of energy ensures sustainable development in any society, preserving valuable energy resources and managing their consumption properly is one of the most important issues on the agenda of all countries in the world, and it has prompted all policymakers, government officials, and energy sector stakeholders to find a solution to the problem of excessive consumption and, as a result, control energy costs.

One of the most important issues that can control the entire cycle of energy production, distribution, and consumption and use these precious resources in the best possible way is energy management, which is today the most important tool for

DOI: 10.1201/9781003740148-11

dealing with the excessive increase in consumption and preventing waste, so that its establishment and application involves optimizing consumption and means choosing the right model of policies in energy consumption, which in addition to being a guarantee for continued economic growth, reduce the destruction of energy resources and also reduce the adverse effects of its improper use on the environment and society.

Energy management includes planning and operation of units related to energy production and consumption, which is the balance between production and consumption. The goals of energy management include resource conservation, climate protection, and cost savings, while users have constant access to the energy they need. Energy management is examined from three perspectives: a strategic perspective, an energy system application perspective, and an alternative revenue perspective. Thus, energy management is considered a key factor for ensuring and maintaining the performance of organizations. The solution to address questions such as how to reduce calculations in the optimization phase and how to increase efficiency are proposed as follows:

- What activities and operations does the effective and efficient process of implementing this type of system involve?
- What information sources are used at each stage of optimization?
- What type of feedback is required at each stage of the process?
- Who has this information and what are the ways to share, integrate, transfer, and use this information?

This research attempts to provide a decision support system for managing and optimizing electrical energy using reinforcement learning. In this research, to deal with uncertainty, a set of different energy consumption states and a set of decisions that can be made are considered in a fuzzy form, and then fuzzy rules for transition states from different states to each other and a scoring system are determined by optimization algorithms, and finally rules are provided to describe what the system can observe.

Today, not only development but also the continuation of life requires energy. In today's world, this energy is supplied through various carriers such as oil, gas, oil and gas products, coal, electricity, and renewable energies. Energy supply in a country like Iran, with a vast geography and different physical locations such as altitude, climate, etc., requires precise and scientific planning.

However, in practice, the country's energy supply in recent years has been planned and implemented without considering optimal energy management considerations and with regard to the ease of access to fossil carriers in the country, which has not only led to the correct choice but also contradicts and conflicts with the growth and development of the national economy. Therefore, the study of energy strategy has been managed in the form of production, distribution, and consumption programs of various energy carriers and providing solutions for optimal production and distribution of energy and consumption.

Sustainable development is an organizing element that ensures the sustainability of non-renewable resources, limited resources that are essential for the life of future generations on Earth. In other words, it means creating a balance between development and the environment. Sustainable development is a process of changing the use of resources, directing investments, directing technological development, and institutional change that is compatible with present and future needs. This sustainability can have four aspects: sustainability in natural resources, political sustainability, social sustainability, and economic sustainability. In fact, sustainable development pays attention to environmental, social, and economic aspects and is the intersection of society, economy, and environment. Sustainable development offers solutions to the structural, social, and economic patterns of development in order to prevent the emergence of problems such as the destruction of natural resources, the destruction of biological systems, pollution, climate change, and excessive population growth. Therefore, the relationship between energy and the environment, or sustainable energy development, is organizing energy sector activities in a way that human development is provided in the short, medium, and long term in all different dimensions.

In this study, we propose a systematic model for energy management and optimization based on reinforcement learning, with the following objectives:

- Extracting energy resources at lower cost and greater efficiency
- Diversifying energy sources
- Creating a downward trend in the environmental impacts of the energy sector
- Creating balance and justice in public access to energy at a reasonable price
- Changing consumption trends
- Changing consumption mix
- Increasing energy productivity and efficiency
- Liberalization and privatization

11.2 BACKGROUND

Researchers proposed a method entitled "Energy Efficiency and Management Analysis Based on Fuzzy Data Envelopment Analysis Model for Ethylene Production Systems in Chemical Industries", an energy efficiency and optimization analysis method based on a fuzzy cross-sectional model with fuzzy data. The proposed method had more power in solving decision-making problems. First, the minimum, median, and maximum values of ethylene energy consumption data in multiple criteria were obtained using data fuzzification. Then, based on the multi-criteria fuzzy data, the criterion of effective production conditions is obtained by the fuzzy data envelopment analysis model. Experimental results show that the proposed method can improve energy production conditions and its consumption efficiency in chemical industries (Fazlollahtabar, 2010).

A research discussed and analyzed an energy management system developed from a typical microgrid under the title "Energy Management and Optimization in a Microgrid System Based on Green Energy". They also considered nonlinear costs and diffusion functions as objective functions. The proposed cost function includes the costs of buying/selling from/to the main grid. The proposed optimization problem is solved using the CPLEX solver based on AIMMS. Research activities are particularly focused on the integration of distributed energy resources and energy storage systems in microgrid applications. In this chapter, the coordination between the microgrid and the main grid for the required variable load demands is performed using an energy management system. For this purpose, an advanced energy management system is introduced in a typical microgrid operating in grid-connected mode. The advanced energy management system is able to determine optimal operating strategies that minimize energy costs, pollutant emissions, and, consequently, the output of available renewable energy sources.

Today, energy conservation and emission reduction in this industry are required to meet the challenges of sustainable development. In complex petrochemical industries, energy optimization and forecasting are effective tools to achieve economic and environmental goals. The ethylene industry is one of the most important sectors of the petrochemical industry. According to statistical data, the ethylene production of China Petrochemical Corporation was 10,420, and the average fuel and electricity consumption (converted to standard oil) was 571.39 kg/tonne of ethylene in 2014. The ethylene production capacity and the average fuel and electricity consumption of China National Petroleum Corporation were 4,976 and 616.7 kg/tonne of ethylene, respectively, in 2014. The energy efficiency of ethylene industries in China is significantly lower than that of developed countries, and there is a great room for improvement in the energy efficiency of China's ethylene industries. Therefore, the study of energy optimization and forecasting of ethylene industries is beneficial to the sustainable development of China's economy. At present, the average method and the optimal index method are commonly used by enterprises to analyze energy efficiency. Since energy efficiency values include indicators and influencing factors, and different indicators have different effects on energy efficiency factors, these methods cannot provide excellent energy efficiency criteria of indicators or factors for optimizing the practical mode of energy use. Since complex petrochemical data have multidimensional characteristics, uncertainty, and noise, it is difficult to accurately determine and predict the energy consumption of complex petrochemical systems. Researchers proposed a data envelopment analysis method for a combined artificial neural network method, entitled "Energy Optimization and Forecasting of Complex Petrochemical Industries Using an Artificial Neural Network Approach". The ANN algorithm is a learning algorithm used for multi-layer neural networks. The proposed approach uses a DEA model with auxiliary variables for sensitivity analysis to determine the effective decision-making units. Finally, the proposed model is validated through an application in a complex ethylene production system in China's petrochemical industry. Optimization results and prediction values are obtained to reduce the energy consumption of the ethylene production system, guide ethylene production, and improve energy efficiency (Hedayati et al., 2014).

One of the effective ways to improve the energy efficiency of complex chemical processes is to establish an energy optimization and analysis model. In this chapter, an energy analysis and optimization modeling based on integrated machine learning is proposed. The proposed model is effectively investigated through chemical processes, and as a result, an energy analysis and optimization framework of chemical processes is obtained. This method is efficient in optimizing industrial production of complex chemicals. Energy optimization and analysis of complex chemical processes play an important role in the sustainable development trend.

Another research introduced a method called energy optimization and analysis modeling based on integrated machine learning analysis in application to complex chemical processes, an energy analysis and optimization modeling based on integrated machine learning. The proposed model was effectively investigated through chemical processes, and as a result, an energy analysis and optimization framework of chemical processes was obtained. At first, index analysis and decomposition were used to decompose the one-dimensional data into three energy performance indices of activity effect, structure effect, and intensity, and then these indices and the production/conductivity of the chemical process were defined as the inputs and outputs of the learning machine, respectively, to build the energy optimization and analysis model. Finally, the proposed method was applied to optimize and analyze the energy status of the ethylene system (Alikhani and Fazlollahtabar, 2014). Nowadays, environmental protection and energy emission reduction have become issues in society. The industrialization of a country is mainly affected by chemical productivity, especially the ethylene productivity of complex chemical industries. The current demand for ethylene production reaches 155 million tons per year and is still increasing. Therefore, energy saving analysis and management are effective ways to improve the support. Energy conservation and management of complex chemical processes play a vital role in the sustainable development process (Fazlollahtabar, 2012c).

The sustainable development strategy has played an important role in building a prosperous society in an all-round way. However, China is a country with a large population, scarce natural resources, and relatively backward conditions in its economy and technology. Therefore, a virtuous circle of social and economic development can be achieved through rational utilization of resources and environmental protection, especially in the complex petrochemical industry. The ethylene industry is the most important sector of the petrochemical industry, in which environmental and economic goals can be achieved by forecasting energy production and management. However, in 2015, the ethylene production and average fuel and electricity consumption of China Petrochemical Corporation were 11005.2/a and 559.06 kg/t ethylene, respectively. The ethylene production and average fuel and electricity consumption of China National Petroleum Corporation in 2015 were 5032 and 594 kg/t ethylene, respectively. The energy efficiency of ethylene production in China is much lower than that of international countries. In addition, more than 50% of the operating costs in ethylene plants come from the energy consumption cost of ethylene plants.

11.3 METHODOLOGY

In recent years, due to the complexity and changes in society, the energy factor has played a fundamental role in the economy and politics of each country, and energy management and optimization, as well as adopting a strategy appropriate to it, are among the main pillars of maintaining the stability and political and economic power of each country. Energy management refers to a set of methods and operations that are carried out in various systems in order to effectively exploit existing energy resources with the aim of using energy properly, maximizing profits, and minimizing costs. In Iran, the energy consumed by many existing industries is three times the pattern of the previous decade, which means that it is impossible to compete in the global market for goods produced in various industries. Energy management is a strategy and should include terms such as optimization and saving. Energy management covers related concepts and is in fact the selection of the appropriate type of energy, i.e., the use of the most economical energy source according to the type of consumption. Inefficient use of energy is considered a waste of energy. Energy-related costs often constitute a small part of the total costs. In other words, energy management is a tool for measuring the level of accuracy in implementing the main activities of energy management. In recent years, the consumption of fossil fuels has been decreasing, and as a result, we will face the depletion of their resources and reserves. On the other hand, energy consumption, especially fossil fuels, is the most important factor in air pollution and climate change. For this reason, the optimal use of energy in the process of economic development has always been considered an important goal in sustainable development. This is possible by optimizing and improving the methods of exploiting resources and energy conversion and transmission processes, as well as implementing energy management by a system appropriate to it. Achieving energy management goals will involve acquiring new equipment or technologies and applying the optimal level of energy throughout the organization, which ultimately leads to minimizing energy costs, waste without affecting quality and production, and minimizing environmental impacts. On the other hand, in order to preserve non-renewable energy resources and their optimal use, which also belong to future generations, society should be diligent in preserving them. One of the reasons why we need an electrical energy management system is to increase energy reliability and minimize power outages. In this research, we design a decision support system based on reinforcement learning for intelligent management and optimization of electrical energy. Due to fluctuations that may exist during electrical energy consumption, energy consumption data is not necessarily definitive. For this reason, the electrical energy consumption management system requires a computational system. On the other hand, in order to simplify decision-making, electrical energy consumption data must be fuzzy, for which we use triangular fuzzification. Also, fuzzy rules are formed for the relevant calculations, and we analyze these rules with the reinforcement learning tool and the Q-learning algorithm. As a result, the present research has three phases: electricity management and optimization, decision support system, and fuzzy rule analysis tool (reinforcement learning) (Fazlollahtabar and Ahmadiahangar, 2024).

The steps and stages of conducting this research to achieve the desired goal are given below.

The first step is to obtain the electricity consumption data, which is considered the input of the system. In this study, we have 1000 electricity consumption data points, which we divided into ten groups of 1000. In the next step, we apply triangular fuzzification to the data using the relevant formulas shown in the next section. One of the reasons we used triangular fuzzification is that the fluctuations of electricity data are high, so we consider their fuzzification to be appropriate for the fluctuation of electricity consumption in different time periods. The next step is to calculate the consumption forecast using the Q-learning algorithm. Reinforcement learning is a solution to the Markov decision problem. This method can find the optimal answer even in the absence of an environment model that includes a transition function and a reward function. One of the most famous reinforcement learning algorithms is Q-learning. The reason for the popularity of this method is the existence of convergence proofs, its apolitically, simplicity, and its suitable convergence speed. In this study, we used this algorithm for the relevant calculations.

The next step, the Markov decision process, is the basic formulation for reinforcement learning. In this formulation, the environment is modeled with a set of states and actions, and the goal is to control the system through research maximization. The decision-making process using the Markov model is a discrete-time random control process. When the outcome of a decision is somewhat random, the Markov decision-making rules are applied. Markov says that without having a large set of data, we can predict the future. In other words, in the model, given the current state, we predict the state of the next period.

The final step is to analyze and interpret the results obtained with the help of fuzzy rules. That is, with the help of if-then rules, we make inferences for decision-making in different consumption states.

11.3.1 Fuzzy Numbers and Operations

When mathematical methods are used, precise rules are required to solve existing problems; however, many descriptions or concepts in the real world are not clearly defined. In a daily conversation, many vague words are used, for example, the facade of this building is beautiful, or the value of the dollar is relatively high. Fuzzy sets are presented to deal with these imprecise words and statements. Fuzzy sets are faced with imprecise concepts, such as sets of tall people that cannot be expressed by regular sets. In the previous sentences, the words tall and close are imprecise. It is not possible to express these expressions with regular sets, and we must express the expressions precisely, such as the set of people who are more than 160 centimeters tall. Measuring the height of a person does not determine whether or not they belong to the aforementioned fuzzy sets. These regular sets that are expressed precisely are known as definite sets in the theory of fuzzy sets. Fuzzy set theory is an extension of definite set theory. We can consider a set in terms of its elements along with the degree of membership of each element.

Let X be an arbitrary reference set. The indicator function of any regular subset A of X is a function from X to [0, 1], defined as:

$$X_A(x) = \begin{cases} 1 & x \in A \\ 0 & x \notin A \end{cases} \tag{11.1}$$

Now, if we extend the range of the indicator function from the two-membered set {0,1} to the interval [0,1], we have a function that assigns to each x of X a number from the interval [0,1]. We call this function the membership function of A. Now A is no longer a regular set but what we call a fuzzy set. Thus, a fuzzy set A is a set whose membership degree can be chosen continuously from I = [0,1]. This set is completely and uniquely characterized by a membership function; a function that assigns to each element of X a number from the interval [0,1] as the membership degree of that element in the fuzzy set A.

A triangular fuzzy number is a fuzzy number represented by three real numbers in the form F = (l, m, u). The upper bound, represented by u, is the maximum value that the fuzzy number F can take. The lower bound, represented by l, is the minimum value that the fuzzy number F can take. The value m is the most probable value of a fuzzy number. The membership function of a triangular fuzzy number is expressed as the following formula:

$$\mu_f(x) = \begin{cases} \dfrac{x-l}{m-l} & l \prec x \prec m \\ \dfrac{u-x}{u-m} & m \prec x \prec u \\ 0 & Otherwise \end{cases} \tag{11.2}$$

According to the membership function of triangular numbers, it is clear that if x is between l and m, then the larger it is, the greater its degree of membership will be, until for x = m the degree of membership will be equal to one. If x is between m and u, the larger it is, the smaller its degree of membership will be, and at x = u the degree of membership will be zero.

The next step is data fuzzification. In this study, we considered a data set from x1 to x1000, which is divided into ten categories of 100. We also obtain the maximum, average, and minimum values of the data using the following formulas:

$$\begin{aligned} D &= \min\{x_i, x_{i+1}, \ldots, x_{i+l-1}\} \\ U &= \max\{x_i, x_{i+1}, \ldots, x_{i+l-1}\} \\ M &= \frac{\{x_i, x_{i+1}, \ldots, x_{i+l-1}\}}{l} \end{aligned} \tag{11.3}$$

$$D' = \begin{cases} M - 2(M-D), M \geq 2(M-D) \\ 0, M \leq 2(M-D) \end{cases}, \quad U' = M + 2(U - M) \tag{11.4}$$

$$\mu_D = \begin{cases} 1, x \leq D' \\ e^{-\frac{1}{2}\left(\frac{\frac{x-D'}{U'-D'}}{\sigma}\right)}, D' \prec x \prec \dfrac{D'+U'}{2} \\ 0, x \geq \dfrac{D'+U'}{2} \end{cases} \tag{11.5}$$

$$\mu_M = \begin{cases} 0, x \leq D' \\ e^{-\frac{1}{2}\left(\frac{x-\frac{D'+U'}{2}}{\frac{U'-D'}{\sigma}}\right)^2}, D' \prec x \prec U' \\ 0, x \geq U' \end{cases} \tag{11.6}$$

$$\mu_U = \begin{cases} 0, x \leq \frac{D'+U'}{2} \\ e^{-\frac{1}{2}\left(\frac{x-U'}{\frac{U'-D'}{\sigma}}\right)^2}, \frac{D'+U'}{2} \prec x \prec U' \\ 0, x \geq U' \end{cases} \tag{11.7}$$

11.3.2 Reinforcement Learning

In this section, we will state the reinforcement learning problem. Since all methods for solving this problem are considered reinforcement learning methods, this problem defines the reinforcement learning branch. Also, in this section, we will state the ideal mathematical state of the reinforcement learning problem. At the end of this section, we will refer to the relationships of its main elements, such as value functions and the Bellman equation.

The reinforcement learning problem creates a suitable framework for learning while interacting and achieving the goal. The learner or decision-maker is called the agent, and what it interacts with, including everything outside the agent, is called the environment. This interaction takes place continuously, and the agent chooses an action, and the environment responds to it and places the agent in a new state. The environment also rewards the agent's actions. The reward is a numerical value that the agent must maximize over time.

The agent and the environment interact with each other in discrete time intervals t = 0,1,2,… and in each time interval t, the agent receives a state (st) from the environment where S is all possible states of the environment and based on that selects an action (at) where A(st) is the set of actions that are available in state st. In the next

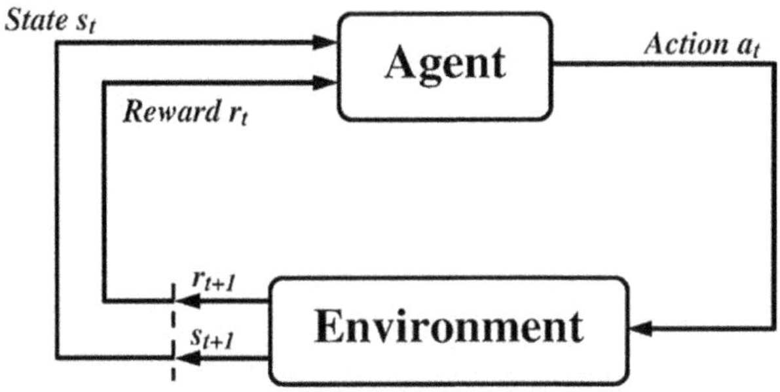

FIGURE 11.1 Agent–environment interaction in reinforcement learning.

time interval and as a result of its action, the agent receives a numerical value rt+1 and is placed in a new state st+1. Figure 11.1 shows how the agent and the environment interact.

At each time interval, the agent makes a map of states with the probability of choosing each possible action. This map is called the agent's policy. Reinforcement learning methods examine how the agent's policy changes as experience is gained.

In reinforcement learning, the agent's goal is formulated based on the reward signal it receives from the environment. At each time interval, the reward is a real number rt, and the agent's goal is to maximize the sum of rewards over time, which means that it does not maximize the reward, but rather the sum of rewards received over time.

The use of a reinforcement signal is a distinctive feature of reinforcement learning. At first glance, this formulation may seem restrictive, but in practice it has been proven to be quite flexible. If we want the agent to solve a problem, we must define the reward in such a way that its maximization over time will lead to the solution of the problem, so determining the appropriate type of reward for solving the problem is absolutely necessary. Another point of reward is that reinforcement signals should not transfer our knowledge to the agent, nor should they specify how to achieve the goal, but should only indicate the goal and what the agent wants to achieve. Another point is that the calculation of the reward takes place outside the agent and in the environment. In other words, the reward is outside the agent's control, and the agent should not be able to change this signal. Of course, the agent can define internal rewards for itself.

It was said that the agent's goal is to maximize the total reward received over time, but how do we express it in a formula? If we define the sequence of rewards received after a time interval t as rt+1, rt+2, etc., in general, we want to maximize the expected return. Return has various definitions, the simplest of which is:

$$Rt = rt+1+ rt + 2 +\ldots+ rt \tag{11.8}$$

where t is the final time interval of the episode. This definition is widely used in practical applications. Because usually the relationship between the agent and the environment is divided into subsequences of intervals called episodes. Each episode ends in a specific state called the final state, and the beginning of the next episode is in a standard starting state.

The rate of decline is another concept that needs to be defined. If the discount rate is used in the definition of the return, the agent chooses actions that maximize the sum of the discounted rewards it receives in the future. In this case, the agent's return relationship is defined as follows:

$$R_t = r_{t+1} + \gamma\, r_{t+2} + \gamma^2\, r_{t+3} + \ldots \tag{11.9}$$

The parameter γ is the discount rate. The discount rate determines the current value of the future rewards.

11.3.2.1 Q-Learning Algorithm

Q-learning is a reinforcement learning technique that follows a specific policy for performing different actions in different situations by learning an action/value function. One of the strengths of this method is the ability to learn the function without having a specific model of the environment.

The first step is to initialize the Q values. The Q table is a table in which the maximum expected future reward for each action in each situation is calculated. Using this table, it is determined what the best possible action is for each state. An example of this table is shown in Table 11.1. In this study, the table Q is the same 1000 electricity consumption data that are considered as Qs of the current period. The next step is to select an action "a" in the current state "s" based on the estimated current Q-value. In the next step, the action selected in step 3 is performed, and the execution of this action outputs state "s" and reward "r". Then, the Bellman equation is used to update Q(s,a). Here, Q(s,a) is defined as the value/action function. The value function also specifies the summation of the accumulated discounted rewards into the future.

The problem model consists of an agent, states (S), and a set of actions (A) for each state. In the relevant research, the agent is a user who consumes electricity at different times. Also, states are ten groups from which energy consumption values for 100 consumption zones were extracted, and actions are the electricity consumption data that are considered as current Qs and are selected step by step to obtain Qs for the next period. By performing an action, the agent moves from one state to the next state and in each state gives a reward to the agent. The agent's goal is to maximize the total reward it receives. This is done by learning the optimal action for each state. The algorithm has a function that calculates the state/action combination. Before learning

TABLE 11.1
Q Table

	S_1	S_2	…	S_m
a_1	$Q(S_{1,} a_1)$	$Q(S_{2,} a_1)$	…	$Q(S_{m,} a_1)$
a_2	$Q(S_{1,} a_2)$	$Q(S_{2,} a_2)$	…	$Q(S_{m,} a_2)$
…	…	…	…	…
a_n	$Q(S_{1,} a_n)$	$Q(S_{2,} a_n)$	…	$Q(S_{m,} a_n)$

starts, Q returns a fixed value chosen by the designer. Then, each time the agent is rewarded, new values are calculated for each state/action combination. The core of the algorithm consists of a simple iterative update, in which the previous values are modified based on new information.

An algorithm episode ends when the agent reaches the final state, the goal.

The pseudocode of the Q-learning algorithm consists of the following steps:

- Initialize the Q values with arbitrary values for all pairs of state and action.
- For life or until learning stops.
- Perform action (a) in the current state (s) of the world based on the current estimate of the Q-value.
- Perform action (a) and observe the output state (s') and reward (r).
- Update Q(s,a).

As we said, to update Q(s,a), we use the Bellman equation, the formula and parameters of which are shown below:

$$NewQ(s,a) = Q(s,a) + \alpha[R(s,a) + \gamma \max Q'(s',a') - Q(s,a)] \tag{11.10}$$

Learning rate ($\propto$): This rate determines how much new information is preferred over old information. A value of zero causes the agent to learn nothing, and a value of one causes the agent to consider only new information. In this study, we set the learning rate to 0.05.

Discount rate (γ): The discount rate determines the importance of future rewards. A value of zero causes the agent to be opportunistic and only considers current rewards. While a value of one encourages the agent to fight for rewards for a long period of time. We also set the discount rate to 0.9 in this study.

R(s, a): The reward for performing the action in that state

Max $Q'(s', a')$: The maximum predicted future reward given the new s' and all possible actions in that new state

New Q(s, a): The new Q-value for that state and action

11.3.2.2 Markov Property

Many decision problems are found in stochastic environments, where the uncertainty of the state of the environment increases the complexity of their analysis and creates very complex problems. A subset of these stochastic problems can be formulated as Markov decision problems. It is appropriate that the state signal summarizes all past perceptions and stores all information about them. To achieve this, more than one direct perception is needed. Of course, this requirement does not go beyond the totality of past perceptions. In the framework of reinforcement learning tools, the agent makes its decisions as a function of the environmental signal, called the state of the environment. Therefore, a state signal that succeeds in storing all past decision information has the Markov property. For example, the position of chess pieces has the Markov property because it summarizes the sequence of all moves that have reached this state. Although most of the information that led to this state is lost, all the important points for the continuation of the game are preserved. This property is sometimes called the path-independent property, which means that the current state is independent of the path and history of past signals.

Assuming a finite number of states and reward values, we give the mathematical definition of the Markov property. In general, an environment at time t + 1 responds to the action at time t, and in the most general case, this response may be related to all past responses. In this case, we characterize the environment by the following complete probability distribution:

For all r and s and all possible values of past events: st, at, rt, st-1, at-1, …, r1, s0, a0. But in the case where the state signal has the Markov property, the response of the environment at time t+1 depends on the state and action at time t.

The state signal has the Markov property if the two relationships above are equivalent to each other for all values of s and r and the past events st, at, rt, st-1, at-1, …, r1, s0, a0. In this case, it is said that the environment has the Markov property. Environments that have Markov properties simplify the problem. Even for a state signal that does not have Markov properties, we can assume that the signal has an approximation of the Markov property.

$$\left|\begin{array}{l}\Pr\{s_{t+1}=s',r_{t+1}=r\mid s_t,a_t,r_t,s_{t-1},a_{t-1},\mathrm{K},r_1,s_0,a_0\}=\\ \Pr\{s_{r+1}=s',r_{t+1}=r\mid s_{t,}\ a_t\}\\ \text{for all } s',r,\text{and histories } s_t,a_t,r_t,s_{t-1},a_{t-1},\mathrm{K},r_1,s_0,a_0.\end{array}\right| \tag{11.11}$$

In a Markov decision process, an agent or decision-maker can influence the state of a system by performing an action that causes the system to optimize a predefined efficiency. To do this, the agent observes the state of the system at specific points in time, called the decision origin, and collects the necessary information to select an operation that will achieve its desired efficiency goal. Each action represents a reward or penalty, and this action affects the state of the system and, consequently, future operations. By applying the selected action to the system, the agent incurs a cost, and the system moves to a new state based on the probability distribution.

An example of a reinforcement learning problem that has the Markov property is called a Markov decision process. If the space of states and actions is finite, then the Markov decision process is finite. A finite Markov decision process is very important for understanding the reinforcement learning tool. A finite Markov decision process is given by a set of states and their actions at a later stage of the environment. For any given state and action s and r, the probability of transitioning to the next state *s'* is:

$$\Pr\{s_{t+1} = s'a' \mid s_t = s,\ a_t = a\} \tag{11.12}$$

A Markov decision process is defined as a manifold M = < S, A, T, R>, where S is the set of states, A is the set of actions, T is the transition function, and R is the reward. The set of behaviors is also denoted by B1, B2, …, Bn.

The transition function represents the probability of going from one state to another and is defined as A × S → [0,1] T: S × and the reward function is the value of being in a state or the value of being in a state and action, which is denoted as R: S → R or R: S, A → R. Finding a solution to an MDP problem can involve finding a policy that maximizes the expected value of the reward over time. A policy is defined as π: S → A.

For the agent, n behaviors are considered, and for each behavior a separate reward signal is given. Therefore, reward signals R1, R2, …, Rn are given to the behaviors. Each behavior also has a state space corresponding to the environment in which it is located, and there are n subspaces S1, S2, …, Sn, one for each behavior. Therefore, for each behavior, a process is considered, which is represented as (Si, A, Ti, Ri). The action space is equal to the action space in the general Markov decision process, and the transition function is also specified over the general process.

All reinforcement learning algorithms are based on the estimation of value functions. Value functions are functions of states (or pairs of state-actions) that estimate the goodness of an agent in a particular state (or the goodness of performing a given action in a particular state). We define goodness in terms of future rewards. It is clear that the rewards that the agent expects to receive in the future depend on the current action. Therefore, value functions are defined in terms of chosen policies.

π is a mapping from state s ∈ S and action a ∈ A(s) to the probability of choosing action a in state s or π(s, a). The value of state s in policy π, V^π (s), is the expected return if the initial state is s and policy π is chosen later in the episode. The state-value function for a Markov decision process is defined as:

$$V^{\pi}(s) = E_{\pi}\{R_t \mid s_t = s\} = E_{\pi}\left\{\sum_{0}^{T} \gamma^k r_{t+k+1} \mid s_t = s\right\} \tag{11.13}$$

where Eπ {} denotes the expected return if policy π is adopted. Note that the values of the final states, if any, are zero. We call V^π the state-value function for policy π.

Similarly, $Q^{\pi}(s,a)$ is the value of choosing action a in state s if we adopt policy π. In fact, $Q^{\pi}(s,a)$ is the mathematical expectation of the return if we start from state s and choose action a and then adopt policy π.

$$Q^{\pi}(s,a) = E_{\pi}\{R_t \mid s_t = s, a_t = a\} = E_{\pi}\left\{\sum_{0}^{T} \gamma^k r_{t+k+1} \mid s_t = s, a_t = a\right\} \tag{11.14}$$

We call Q^{π} the action value function for policy π.

The value functions V^{π} and Q^{π} can be estimated using experience. For example, suppose that an agent adopts policy π and keeps an average of the actual returns it receives after each state it encounters. As the number of times the state is observed increases, this average will converge to the state value V^{π}(s). If separate averages are kept for each state-action pair, these averages will converge to the action values Q^{π}

(s, a). Such estimation methods are called Monte Carlo methods. Of course, if the number of states increases, then it is not practical to keep an average for each state independently. Instead, we express V^{π} and Q^{π} as parametric functions and determine the parameters in such a way that they best match the observed returns.

Having a recurrence relation is a fundamental property of value functions. For any policy π and any state s, the following relation exists between the value of state s and the value of its subsequent possible states:

$$
\begin{aligned}
V^{\pi}(s) &= E\{R_t \mid s_t = s\} \\
&= E_{\pi}\{\sum_{0}^{T} \gamma^k r_{t+k+1} \mid s_t = s\} \\
&= E_{\pi}\{r_{t+1} + \gamma \sum_{0}^{T} \gamma^k r_{t+k+2} \mid s_t = s\} \\
&= \sum_{a} \pi(s,a) \sum_{s'} P^a_{ss'} [R^a_{ss'} + \gamma E_{\pi}\{\sum_{0}^{T} \gamma^k r_{t+k+2} \mid s_{t+1} = s'\}] \\
&= \sum_{a} \pi(s,a) \sum_{s'} P^a_{ss'} [R^a_{ss'} + \gamma V^{\pi}(s')]
\end{aligned}
\tag{11.15}
$$

11.4 COMPUTATIONAL IMPLEMENTATIONS AND FINDINGS

To implement the proposed model, we use the electricity energy consumption data of manufacturing industries in the industrial towns of Mazandaran Province. This information was obtained from the Mazandaran Province Electricity Distribution Company. This case study presents an analysis of the electricity energy consumption data of manufacturing industries in Mazandaran Province with regard to regional changes.

Due to the high consumption of electrical energy among the types of energy, the management of the consumption of this type of energy is of great interest. Small consumers such as residential homes have smaller electrical power consumption compared to large industrial consumers, but due to the large number of domestic consumers, they account for a major share of the grid power, so it is necessary for each residential home to manage its electrical energy to ultimately reduce consumer costs, prevent energy waste, and increase the stability of the electrical grid. A reliable and efficient power grid plays a major role in a country's infrastructure and has a great impact on its economic and social progress. Uninterrupted, safe, and efficient flow of electricity from generators to consumers is not only desirable but also a major requirement for a reliable power grid, where electricity must be generated and delivered on time. The sustainable development of the region depends heavily on the health of renewable resources such as soil, water, vegetation, livestock, etc. Therefore, the integrated development of all these structures is essential for environmentally friendly development. Currently, the focus of energy sector planning is from demand analysis to energy demand dynamics management and conservation with a good understanding of the factors affecting the growth and pattern of energy demand before entering into its forecasting. Demand analysis involves the collection and presentation of a consistent set of information on consumption and demand for different types of energy.

Electrical energy analysis for different fuels can be achieved with an efficient decision support system (DSS). A decision support system provides an integrated framework for easy access to data analysis and design and evaluation of domestic energy consumption strategy with a user interface. A decision support system helps in the analysis of energy resources. The technologies and methods used to develop and use DSS to assist in domestic energy consumption make the task easier for a decision-maker. It also provides a great advancement in the ability to quickly access and process large databases at high speeds. Therefore, designing a decision support system for energy management and optimization improves the quality of decision-making at the block and regional levels and provides analysis and evaluation of the effects of various decisions.

In the previous sections, we examined the components of a decision support system. The decision support system consists of three parts: input, processing, and output. In this study, the input includes electrical energy consumption data. In the computational core, we use the Q-learning algorithm with fuzzy data and interpret and analyze its output as the DSS output.

Based on the architecture of the previous figure, the design of a decision support system for managing and optimizing electrical energy includes the following steps:

(1) The first step in the DSS development process is the design of the user interface, in which special attention is paid to the user visualization tools, such as maps and charts, reports, etc., which are valuable for helping the decision-maker (energy expert in the electric company) in decision-making. The main goal of the user is to increase work productivity, reduce operating costs, and response time. The term user interface covers all aspects of communication between the user (decision-maker) and the decision support system. This system includes not only hardware and software but also agents and tools with easy accessibility and use. The user interface is a system from the perspective of the manager or user because it is the only part of the system that the user sees and has access to. Also, one of the reasons why users cannot use quantitative analysis and computers is the hard user interface. In this study, the user interface is a system that records electricity consumption and has editing, file, and reporting tools.

(2) The next step is to design the data management subsystem. In this architecture, the relevant subsystem includes an electricity database that contains electricity data (1000 data). This data is considered as input to the decision support system. The electricity database is managed by software called a database management system (DBMS). One of the most important tasks of this software is to maintain the security and integrity of information. Access to information in a file system requires physical access to the information. This makes it almost impossible to control access to specific data, and in addition, it will not be possible to share the processing load when information needs increase. In this study, the collected electricity consumption data is located in the Excel program.

(3) The next step is the main processing of data and information, which is referred to as the computational core of the decision support system. In this step, we use the Q-learning algorithm and the Markov decision process with fuzzy data. This step is one of the main steps in designing a decision support system in which the system takes input, processes it, and gives the output to the user.
(4) The next step includes analyzing electricity consumption, which analyzes energy consumption based on the level of analysis, period, and time, and finally we interpret the output of the processing using fuzzy rules. That is, with the help of these rules, we make inferences for making decisions in different modes of electricity consumption.

11.4.1 Numerical Results of Implementation

In this section, we examine the numerical results of the implementation in the form of tables and graphs. As we said in the previous sections, our research has 1000 electricity consumption data that are considered as input to the system. We divided the data into ten groups of 100. In other words, we extracted the energy consumption values for 100 consumption ranges. The data set of electricity energy consumption is shown in Table 11.2. According to the Q-learning algorithm, these data are the current Qs. According to the formulas in Section 11.3, we triangularized these data in each group in order.

In the next step, the Qs of the new period are obtained through the Bellman equation. This table shows the electricity consumption data of the future state, so that by updating the equation, the Qs of the subsequent periods are also obtained, and the procedure continues for several steps. In this study, we calculated up to two periods of electricity consumption: one period of the current state and one period of the future state.

In the current and predicted Q table, for each group (ten zones), we calculated the average function. In this way, we calculate the sum, minimum, and standard deviation functions of the two periods for the Qs of each range (ten groups) in the current and predicted state. The results of these functions are the output of the decision support system, which in the next step will be shown in the form of a graph for further review and analysis. Finally, the analysis of the results is provided to the decision-maker for decision-making. Power plants exist as a source of energy production in every region. The goal is to determine patterns in different situations so that we can manage and optimize energy. For example, if the power plant's production capacity is 66083 kWh, the sum of the consumption in some groups is greater than this production rate. So, for these groups, we must determine a saving pattern that does not exceed that limit and can establish a balance.

The data shows the fluctuations in electricity consumption in the current and forecasted periods. For example, if the total energy consumption of a group is 275 kWh and the power plant's generation capacity is 1000 kWh, this power plant should not produce energy at the same rate as it consumes and should use the concept of tolerance to produce energy within the desired energy consumption range. In other words, the power plant should produce electricity according to the obtained interval and

TABLE 11.2
Set of Values of Electricity Consumption for 100 Beats

Group 10	Group 9	Group 8	Group 7	Group 6	Group 5	Group 4	Group 3	Group 2	Group 1
1039	771	663	831	687	701	284	407	311	488
299	208	716	602	1091	221	336	620	338	217
868	729	500	579	286	514	909	926	995	691
897	859	835	560	1097	504	626	292	911	1038
989	986	779	250	654	709	914	423	289	1021
739	1047	948	906	604	430	247	309	549	295
662	284	1026	362	685	1039	864	861	734	456
682	1081	883	7 73	791	522	835	872	613	923
761	965	1095	491	363	769	1018	370	561	1004
345	258	732	510	1042	392	351	324	990	248
219	554	515	530	858	1080	979	926	745	975
814	1084	975	931	1082	551	757	381	585	858
652	491	733	827	789	608	754	365	695	535
1006	383	546	586	963	497	1089	415	901	971
536	268	483	1018	1013	912	536	1013	1063	857
836	457	1035	472	262	1047	710	863	1027	389
752	744	1059	1067	476	420	594	994	1062	630
987	362	622	571	809	696	1099	875	955	438
201	662	843	213	867	501	239	219	311	514
789	359	201	642	493	345	995	970	997	1023
427	778	536	913	250	840	741	853	430	1081
1028	1035	428	516	1014	870	1023	624	349	584
656	668	876	681	200	338	474	827	346	329
1061	758	1047	466	214	859	740	429	518	208
245	729	857	294	660	279	400	643	695	663
476	745	492	377	688	522	1030	968	819	445
316	900	562	349	690	875	859	491	815	884
922	240	757	388	584	942	675	510	1046	646
584	213	790	853	209	614	575	763	262	1035
215	848	447	708	400	1058	279	935	754	434
441	885	520	365	854	560	839	543	741	674
609	934	375	921	435	1013	1091	931	1091	1025
840	343	348	895	710	330	996	514	881	606
808	923	618	772	906	607	659	935	906	1066
445	770	733	1085	802	1067	790	493	835	767
778	335	770	843	418	954	373	1080	547	772
469	387	469	952	373	944	1017	404	670	489
818	473	657	385	949	603	944	330	329	989
531	795	916	665	808	262	549	713	639	358
620	564	791	489	978	813	427	1010	950	1083
567	250	531	652	326	283	661	781	459	930
601	930	705	1018	425	1080	786	709	752	933
709	1042	580	530	471	1029	305	1059	890	1100
361	815	546	460	661	692	359	228	1074	559
235	241	837	330	775	1005	510	672	462	459
224	974	548	513	568	1002	490	437	953	778
964	969	827	347	608	378	362	794	317	529

(Continued)

TABLE 11.2 (Continued)
Set of Values of Electricity Consumption for 100 Beats

Group 10	Group 9	Group 8	Group 7	Group 6	Group 5	Group 4	Group 3	Group 2	Group 1
480	1012	1033	467	802	414	687	289	205	596
781	317	356	1052	912	457	499	757	855	693
1023	248	657	943	524	684	435	481	1058	322
619	486	355	200	496	523	766	1064	702	733
1057	713	694	569	906	995	465	773	548	654
302	1083	553	773	993	937	916	343	302	310
785	735	635	669	359	800	631	269	275	305
1028	1092	403	361	735	1086	812	970	522	435
386	312	816	707	737	484	324	408	441	588
275	203	856	1017	684	804	495	836	751	344
958	234	578	498	277	1072	426	340	515	438
912	850	883	796	881	293	373	1037	697	258
336	702	286	883	250	583	817	227	763	1046
1100	210	1021	390	754	207	761	635	219	411
217	595	770	427	1052	858	944	616	901	557
861	1001	486	736	875	372	333	291	275	621
644	574	962	706	990	733	680	918	679	444
478	771	252	456	306	514	575	415	651	280
655	863	532	945	482	968	1063	592	915	1095
413	412	357	528	917	531	274	422	651	289
299	395	693	1049	240	920	1095	756	209	474
782	766	532	932	602	414	686	1082	217	761
527	296	234	671	655	659	1044	1088	579	862
843	921	940	930	581	550	833	805	313	874
524	1098	893	754	200	259	266	344	999	918
431	208	311	500	426	328	456	227	232	781
1064	868	308	493	581	681	1024	933	1086	934
516	446	768	790	959	1093	361	693	981	299
266	455	668	394	688	504	254	1099	968	423
591	529	315	222	352	965	587	591	392	779
586	876	517	455	767	460	239	1090	486	911
652	1052	422	783	683	735	464	889	440	979
598	673	624	342	410	664	760	705	626	950
836	996	542	377	605	512	887	423	856	476
972	356	539	246	917	909	546	740	345	1009
825	706	685	930	272	374	949	760	458	1006
637	258	1089	1092	219	553	252	321	823	708
468	337	373	670	658	545	959	1031	1062	896
900	464	612	413	714	680	396	672	914	429
810	235	762	1086	427	728	475	1093	220	251
302	947	351	496	249	328	417	590	265	235
649	1053	448	943	611	647	238	246	783	761
602	1015	474	766	373	892	484	351	394	665
751	845	380	445	953	710	230	627	461	382

balance the energy consumption in different time periods. Here, tolerance is the permissible changes in the size and rate of electricity consumption, which is shown as follows:

$$T = \text{Total volume of electricity consumption} \pm \text{power plant generation capacity} \tag{11.16}$$

Now, in order to examine the data in the previous tables more closely, we will show them in the form of a chart. The row of the chart, the ten groups, and the column of the chart show the results of the functions in each group in the current and forecasted periods. As we can see in Figure 11.2, the average energy consumption for each group in the future period is higher than the average energy consumption in the current period, and the amount of electricity consumption in the future will be higher. Over time, the consumption coefficient increases and new units and equipment are added to industrial units, all of which will lead to an increase in electricity consumption in the future.

In Figure 11.3, the total energy consumption for each group in the future period is higher than in the current period. This is because the amount of consumption in the future or the consideration of adding new equipment in industrial units will increase.

As we can see in Figure 11.4, in group 7, the minimum energy consumption of the current period and the predicted period are at the same point. Lack of development, peak consumption, etc. are some of the reasons for this.

Interpreting Results with Fuzzy Rules

Logic is the science that discusses the truth or falsehood of propositions. In classical logic, a proposition is either completely true or completely false. In fuzzy logic, a proposition may be completely true or completely false, or in most cases, partially true and partially false. In other words, the truth value of a proposition can be a number between zero and one. Also, one of the characteristics of fuzzy logic compared to classical mathematics is the use of linguistic variables to solve a problem or to reason

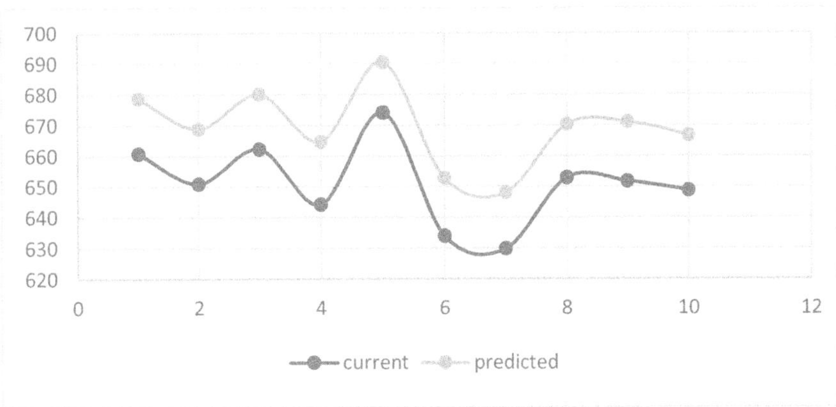

FIGURE 11.2 Average of ten groups in the current and future situation.

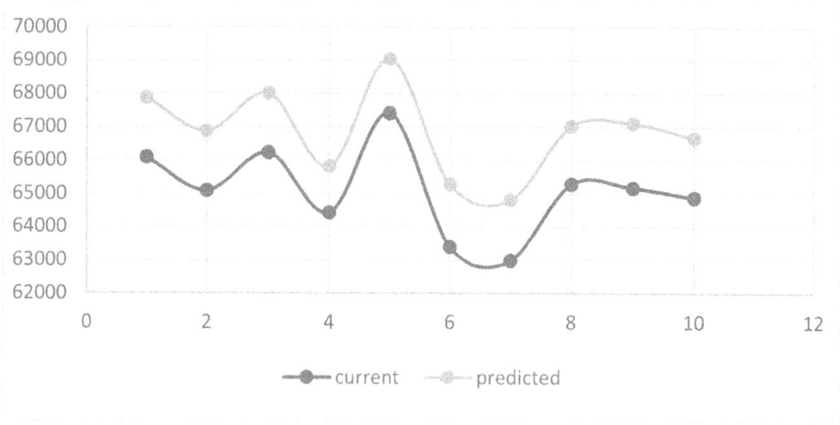

FIGURE 11.3 Sum of ten groups in the current and future situation.

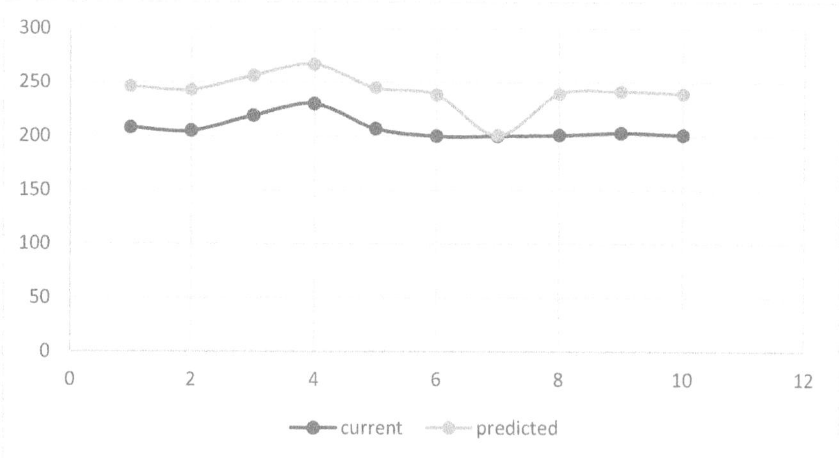

FIGURE 11.4 Minimum of ten groups in the current and future situation.

about a specific situation. In general, if a variable can accept words from natural language as its value, then it is called a linguistic variable. The words are specified by fuzzy sets in the range that the linguistic variables are described by. The most common way of presenting linguistic expressions to express human knowledge and information is to use conditional statements in the form of if-then statements in the following form: If (premise, assumption) then (consequence, conclusion). Consider the following conditional statement:

If x is *A* then y is *B*

In the above conditional statement, A and B are fuzzy. In fuzzy logic, the requirement is that if X is partially true, then Y is equally true. Using the above statement method, we can deduce another fact from a fact that is known to us. This method can well express human empirical and mental information in colloquial language. The antecedent may be a simple or compound fuzzy statement, where the simple statement can be x is A, where x is a linguistic variable, and A is the value of the variable x. A compound statement is a combination of simple statements that are combined using the conjunction or not, which indicates commonality, association, and are fuzzy complements and are connected to each other.

Now, to analyze the output and results obtained in the previous section, we use fuzzy rules. That is, using if-then rules, we make inferences to make decisions in different electricity consumption states:

Rule 1: If the amount of energy consumed ≥ the average amount of electricity consumption, then disconnect the power of user n.

Rule 2: If the amount of energy consumed ≤ the minimum amount of electricity consumption, then encourage user n.

Rule 3: If the amount of energy consumed ≥ the average amount of consumption + the standard deviation of electricity consumption, then the power plant is unable to produce electricity.

Rule 4: If the total amount of energy consumed > the power plant's production capacity, then disconnect the power to user n.

In the previous section, the importance of the research method using the reinforcement learning technique was mentioned. At the beginning of this section, we designed a decision support system for implementation, which includes input, processing, and output components. The input of this system is the electricity consumption data, which we extracted in the form of an Excel table and then performed data fuzzification on them. After that, with the help of the Q-learning algorithm, we obtained the electricity consumption data for the future period. For each of the ten groups in the current and predicted periods, we calculated functions for the behavior and state of the system, which were considered as the numerical results of the research (system output). Then, we showed these results in the form of tables and graphs for further understanding. In the last step, we analyzed these results using fuzzy rules to draw conclusions for decision-making in different energy consumption situations. In the next section, we will discuss the achievements of this research and suggestions for future research.

11.5 CONCLUSION AND SUGGESTIONS

As mentioned, energy management and optimization play a key role in sustainable development. The goal of energy management and optimization is to balance consumption and production. In this study, a decision support system for managing and optimizing electricity was proposed based on reinforcement learning tools. The extracted electricity consumption data was considered as the input of the decision support system. The Q-learning algorithm was considered the computational core of the designed system, and using this algorithm, we calculated the electricity consumption data for the future period. Then, we divided the energy consumption data for the future period into ten groups of hundreds, similar to the current period, and examined the behavior and status of each group in the form of functions. The next step was the system output, which we analyzed using fuzzy rules. Among the features of the reinforcement learning method is the ability to learn through interaction and without the need for a guide. Reinforcement learning, on the one hand, has a strong support of mathematical theorems and proofs, and on the other hand, it has been successful in various problems because the goal of this method is to maximize system performance. The two features of trial-and-error search and delayed reward are the most important features of this technique that distinguish it from other methods. The Q-learning algorithm, which is one of the algorithms of the reinforcement learning technique, converges more slowly than other algorithms of this method, but it has the ability to continue learning as policies change, which is one of the reasons for using this algorithm in the relevant research. Also, the decision support system provides an integrated framework for easy access to data analysis, design, and evaluation of internal energy consumption strategies. For this reason, we used the decision support system to manage and optimize energy and facilitate user decision-making. In this study, the proposed decision support system can adapt to changes in electricity consumption. That is, when the energy consumption of a region changes, the electricity consumption data is input to the Q-learning process, and the output is obtained in real time. In this way, the fluctuations in energy consumption can be managed through the reinforcement learning algorithm.

The implementation results show that the predicted values in the average and total energy consumption situations are slightly higher than the current consumption, which is due to the advancement of technologies and industries, which leads to more electricity consumption in the future. Therefore, more energy consumption in the future seems reasonable. In the standard deviation situation, the predicted values were obtained lower than the current values. The lower the fluctuations in future electricity consumption compared to the current period, the better, as we achieved this goal in this research using the mentioned technique. Also, one of the advantages of the proposed method is that it provides a future predicted value based on the expansion of energy consumption of groups. It is important for policymakers and energy future-minded people to have a reliable trend of consumption for optimal allocation and distribution.

Given the increasing energy consumption in industries and the lack of comprehensive energy consumption standards, laws, and regulations in this field, it is essential to carry out various projects to develop energy consumption standards and manage and optimize various types of energy in various industries. The following are suggestions for future research in this field:

- Using deep learning techniques:
  Deep learning is actually learning by neural networks that have many hidden layers. The further we go in the layers of a deep neural network, the more complex and complete the models we reach. The advantages of this method include automatic feature learning, multi-layer feature learning, high accuracy in results, the potential to create more capabilities and applications in the future.

- Using simulated neural networks in computing:
  In the future, researchers can use the neural network method in the computing core of the system to identify data patterns/trends and perform learning tasks. Neural networks are able to classify inputs to obtain appropriate outputs. Since neural networks are expressed as probabilities, the outputs of neural networks can serve as specific predictions.
- The discussion of electrical energy in industrial units can be raised in renewable energies.
- Forecasting electrical energy production by power plants.

12 Integrated Analytic Network Process (ANP) and Best-Worst Method (BWM) for Online Shopping Security

12.1 INTRODUCTION

In today's growing world, half of the events of our lives take place in the virtual world. Information exchange, receiving and sending video and audio files, confidential information, bank account control, buying and selling goods, and many other things take place in the virtual environment of the Internet. In today's world, the old form of everything has changed; trade is no exception to this rule. The new form of trade is electronic commerce. Researchers have shown that choosing electronic commerce leads to various and numerous advantages, including reducing costs, improving product quality, contacting new customers or suppliers, and creating new ways to sell products. Despite all the advantages that electronic commerce brings, conducting online transactions and communications provides a greater opportunity for misuse of technology and even criminal acts. These problems are not unique to electronic commerce and are part of the widespread problems that are besetting information and computer systems around the world. Every year, many organizations are targeted by security-related crimes (from virus attacks to commercial fraud, such as the theft of sensitive business information and confidential credit card information). Such security attacks cause millions of dollars in losses and disruptions in the activities of companies. The development of information technology applications in the form of the growth of the penetration of the World Wide Web around the world is undeniable, which has had a very positive impact on electronic commerce (Mahdavi et al., 2010).

One of the most important threats and challenges facing this highly used area of information technology is the security of shopping and banking services. Phishing is the most obvious type of attack on this type of service. The theft of identity information of cyberspace users and customers of electronic stores and financial institutions is called phishing. Recently, phishing attacks targeting financial services have become a serious criminal business. One of the new business activities that has emerged in recent years is the creation and development of electronic commerce. Therefore, given the importance of this issue, the present study intends to examine the impact of information security on electronic commerce. One of the necessary and

DOI: 10.1201/9781003740148-12

essential infrastructures for electronic commerce is security, which is considered a very important issue in the world of electronic commerce, and has many challenges. The lack of security in cyberspace can be an obstacle to the growth and development of electronic commerce. In multi-criteria decision-making methods, a number of options are evaluated according to a number of indicators to select the best option. In this method, the best and worst indicators are determined by the decision-maker, and a pairwise comparison is made between each of these two criteria (best and worst) and other indicators. Therefore, the security effects of electronic payment in online purchases are analyzed using BWMs.

Today, we live in a world where information processing is cheap, and communication costs are decreasing, and the world is increasingly exchanging and interacting with information. However, the provision of new technical facilities has not only led to the emergence of new products and better and more efficient ways of doing things but has also increased the possibility of misuse of technology. Information and communication technology, like other technologies, is a tool and can be used in a way that is beneficial to everyone or used in a way that produces dangerous results (Fazlollahtabar and Ashena, 2013).

Most users are non-professionals, and therefore their lack of sufficient information may cause them to not use secure software packages properly, and as a result, hackers and computer criminals can attack and misuse the system regardless of their geographical location or the user. In addition to all these issues, the issue of organized crime in the cyber world adds to the complexity of the work of governments to ensure the security of vital public service infrastructure, and the serious consequences of criminals' misuse of government resources double the importance of addressing the issue of security correctly and effectively. Information security and securing computers have become an undeniable necessity in the information age. Addressing the issue of information security in simple language is felt more than ever before, because each of the human and non-human factors has a defined place in the information security engineering system. A fundamental and logical familiarity with this engineering system and awareness of the elements in this structure, along with scientific knowledge of the responsibility of each of the above elements, is necessary and vital.

The conceptual model is a theoretical assumption based on which the analysis of the security effects of electronic payments in online purchases using best-worst methods is determined. In this process, the elements of the conceptual model are presented in the form of independent and dependent variables. The conceptual model is the analysis of the security effects of electronic payments in online purchases using best-worst methods. The types of variables and indicators of the conceptual model of the research are based on the research literature. The aim of the present study is to examine the security effects of electronic payments in online shopping.

12.2 BACKGROUND

Researchers found that the status of information security in hospital information systems of hospitals affiliated with Tehran University of Medical Sciences and

Shahid Beheshti University of Medical Sciences was at an acceptable level from the perspective of the computer unit managers of these hospitals. However, information security from the managerial and physical aspects in hospitals affiliated with Tehran University of Medical Sciences received a lower score than hospitals affiliated with Shahid Beheshti University of Medical Sciences. In contrast, most hospitals affiliated with Shahid Beheshti University of Medical Sciences had received a lower score than hospitals affiliated with Tehran University of Medical Sciences in terms of the status of information security in hospital information systems from the technical aspect. Therefore, it should be noted that the current situation also has weaknesses, and planning to develop and implement the latest security policies and guidelines in all three dimensions of management, technical, and physical security in accordance with the needs of users and technological advances seems essential (Taheri et al., 2014).

Research works in the literature are described as conducting any business online and via the Internet. This technique has grown significantly in recent years and is expected to grow even more. Electronic commerce is any transaction in which the purchase and sale of goods or services takes place via the Internet and leads to the export and import of goods and services. Electronic commerce usually has a broader application, meaning it not only includes buying and selling via the Internet but also other aspects of business activity, such as purchasing, taking stock of goods, managing production and procurement, distribution, and movement of goods, as well as after-sales services. Of course, the broader concept of e-commerce is e-business (Fazlollahtabar, 2022b).

Identifying and ranking factors affecting information security in electronic banking using fuzzy foundation technique are worked out, concluded that all bank customers and all consumers of electronic banking services are afraid of information security issues, and since the smallest security problem leads to tremendous financial loss and also damage to the image of the main customers of banks, it seems that the development and implementation of security solutions are considered to be the main concerns of such organizations. One of the most important factors in the acceptance of new information technologies by both customers and employees is ensuring information security and preserving the privacy of individuals (Arabshahi et al., 2020).

12.3 METHODOLOGY

If the data required by the researcher are in the range of events in which the researcher must change the facts regarding a variable or, in general, it is possible to observe the variable under study in the future, then the research design is prospective. According to the aforementioned definition, this research is also prospective. The purpose of a series of studies is simply to find an answer to a problem that is not immediately applicable, which is called outcome-oriented research. While in decision-oriented research, the researcher's goal is to find a problem whose results can be immediately used in decision-making, the present study is a result-oriented study. Research is also categorized based on the purpose: applied and fundamental. The purpose of applied research is to acquire the necessary knowledge to determine a tool by which a specific need can be met. The purpose of this type of research is to try to answer a practical problem that has occurred in the real world. According to the definition provided, this research is applied research.

Experimental and descriptive research: Descriptive research includes methods that aim to describe the conditions and phenomena under study and are divided into the following types: survey, correlation, action research, case study, and post-event. Survey research is used to describe the characteristics of a statistical population and is suitable for answering questions such as the nature of existing conditions, the relationship between events, and the current situation.

This research is a descriptive-survey type, because in the first step, local and appropriate criteria of the university were extracted, and then the importance of the criteria was determined based on the opinions of experts and users in this field. The statistical population consists of people who have at least one common attribute, and the researcher studies one of their variable attributes. The statistical population in this research consists of students in the fields of e-commerce and information technology, as well as experts in electronic banking.

Sampling method: In this research, it was important for experts to have knowledge of information security in online shopping; for this purpose, a questionnaire was presented to a number of experts who met the conditions for responding to the questionnaire and also agreed to cooperate in completing the questionnaire. Therefore, data collection in this research was carried out using the convenience sampling method.

In this study, the library method was used to collect information related to the literature on the subject. This method includes the use of books, articles, and magazines, the use of the Internet and related sites, existing documents, and evidence. A questionnaire was used to collect the required data to examine the relationships. A questionnaire is one of the common and direct tools for obtaining research data. A questionnaire is a set of questions to which the respondent must provide the necessary answers with due consideration. These answers are the data required by the researcher.

For this purpose, a questionnaire has been prepared: A questionnaire related to the fuzzy Delphi, in this research, using the library method and reviewing the literature related to the research, as well as with the help of respected supervisors and consultants, factors affecting security were identified and extracted. The general factors identified were provided to university experts for confirmation via email and in person (with necessary explanations), and finally 12 factors were confirmed, which were classified into three categories – technological factors, human factors, and managerial factors. A questionnaire related to identifying the relationships between criteria and sub-criteria, which are specified in this questionnaire according to the experts' opinions (to determine the internal or external dependence of the criteria). The questionnaire is related to the evaluation and pairwise comparisons of clusters. These factors were provided to the company experts in the form of a questionnaire to determine the importance of each factor in the form of a table and against each factor on Saati's nine-point scale to assess the importance of each factor. This questionnaire was distributed to 15 experts to determine the degree of importance of the clusters relative to each other based on the defined relationships; the questionnaire related to

TABLE 12.1
Criteria and Sub-Criteria

No	Criteria	Sub-criteria
1	Management factors	Having sufficient resources
2		Employing strong and competent consultants
3		Cyberspace management
4		Strict control and supervision
5		Support and backing of top management
6		Proper management of financial transactions
7	Technological factors	Internet server equipment
8		Bank
9		Website Network Security
10		Data management (accurate, reliable, sufficient, and appropriate information)
11	Human factors	Perceived knowledge of human factors
12		Perceived security

Best-Worst Method (BWM) includes two tables, one of which is related to the best criterion and the importance score it has compared to other criteria, and the other is related to the worst criterion and the degree of importance that others have on this criterion. Experts determine these degrees of importance from a 1 to 9 range.

Questionnaire specifications:

A. Instructions on how to answer the questionnaire
B. General questions (demographic): age, level of education, work experience, field of study
C. Specialized questions

In this research, in the first stage, criteria and sub-criteria are extracted from library resources and research background review, as well as the opinions of academic experts. In this study, we have 3 main criteria (technology factors, human factors, and management factors) and 22 sub-criteria. In the next step, the important criteria for the organization should be extracted from the point of view of the organization's experts using the fuzzy Delphi method (Table 12.1).

The next step is to obtain the weights of the desired criteria and sub-criteria through the two BWM and ANP methods; of course, for greater accuracy, the output of the ANP method, i.e., the relative weight of the criteria, is used as the input of the BWM method to increase the accuracy of the ranking.

12.3.1 Fuzzy Delphi Method

The Delphi method was first presented in 1963 at the Rand Corporation. This technique is a survey method based on expert opinions and has three main characteristics: anonymous response, controlled repetition and feedback, and finally statistical group response. This technique is a systematic method for collecting and coordinating the informed judgments of a group of experts about a specific question or issue. In many real situations, expert judgment cannot be expressed and interpreted as definite quantitative numbers; in other words, definite data and numbers are insufficient to model real-world systems due to the ambiguity and uncertainty in the judgment of decision-makers. In this regard, in order to overcome this problem, the theory of fuzzy sets, which was presented by Zadeh in 1965, is a suitable tool for dealing with ambiguity and uncertainty in the decision-making process. Therefore, in this study, the fuzzy Delphi method was used to confirm the factors affecting security. This method is a combination of the Delphi method and fuzzy set theory, which was presented in 1993. The steps of the fuzzy Delphi method are as follows:

Step 1: Identifying factors affecting security by comprehensively reviewing the research literature.

Step 2: Collecting the opinions of decision-making experts: In this step, after identifying the factors, a decision-making team consisting of experts related to the research topic is formed and a questionnaire is sent to them to determine the relevance of the identified indicators to the main research topic, in which the linguistic variables of the table are used to express the importance of each indicator. There are different types of fuzzy numbers, such as triangular, trapezoidal, and exponential fuzzy numbers. In this study, triangular fuzzy numbers, which have been repeatedly used by various researchers due to their simplicity in understanding, have been used, where $\tilde{M} = (l,m,u)$ is a triangular fuzzy number and L, m, and u represent the smallest, most probable, and largest possible values, respectively.

Verbal expressions for verifying effective security factors are clarified using linguistic variables and fuzzy numbers as Very low (0,0,0.25), Low (0,0.25,0.5), Medium (0.25,0.5,0.75)

High (0.5,0.75,1) and Very high (0.75,1,1).

Step 3: Verifying important indicators: This is done by comparing the acquired value of each indicator with the threshold value $\tilde{S}$. The threshold value can be determined in several ways, but using the average value of the indicators as the threshold value is one of the most reliable methods. To do this, first the triangular fuzzy values of the experts' opinions must be calculated, then their fuzzy average must be calculated to calculate the average of the opinions of n respondents. For this purpose, it is necessary to calculate the triangular fuzzy number of each indicator. The triangular fuzzy number is created for each of the indicators using the following formulas.

$$\widetilde{a_{ij}} = \left(a_{ij}, b_{ij}, c_{ij}\right) \quad \text{with} \quad i = 1,\ldots,n;\ j = 1,\ldots,m \tag{12.1}$$

$$\widetilde{\tau_{ij}} = \left(a_j, b_j, c_j\right) \tag{12.2}$$

$$a_j = min\{a_{ij}\} \tag{12.3}$$

$$b_j = \left(\prod_{i=1}^{n} b_{ij}\right)^{\frac{1}{n}} \tag{12.4}$$

$$c_j = max\{c_{ij}\} \tag{12.5}$$

In which the index refers to the expert and the index refers to the decision-making index. $\widetilde{a_{ij}}$ is the fuzzy value of each indicator acquired by each decision-maker, and $\widetilde{\tau_j}$ is the fuzzy average of the value of each indicator.

Also, the average of the fuzzy values calculated through Equation (12.6) is defuzzified using the center of gravity method.

$$Crisp = \frac{a+2b+c}{4} \tag{12.6}$$

After calculating the above values, if the de-fuzzy value $\widetilde{\tau_j} \geq \underline{\tilde{s}}$, the desired indicator is approved and enters the main decision-making stage. However, if the de-fuzzy value $\widetilde{\tau_j} < \tilde{S}$, the desired indicator is rejected.

12.3.2 Analytical Network Process

This method was first described by Thomas El-Saati in 1996. This technique combines expert opinions and evaluations, and then a scale-based evaluation method is used to examine the relative importance of pairwise comparisons among each criterion. This technique effectively examines quantitative as well as qualitative indicators.

In fact, ANP is an extended form of AHP. While AHP provides a framework for one-way hierarchical relationships, ANP considers the most complex internal relationships of decision levels and ratios and is very suitable when the dependence between the criteria for selecting possible alternatives is high. In the ANP method, the structure of relationships is networked.

The ANP method is preferred to other multi-criteria decision-making methods for the following reasons:

1. ANP has a systematic approach in determining priorities and weighing between objectives and criteria, and the degree of importance and weight of criteria relative to each other is determined based on the judgments of individuals and not arbitrarily by contract.

2. ANP can consider all visible and invisible criteria in the model.
3. ANP has a relatively simple and understandable approach that is easily accepted by managers and decision-makers.

Steps in the network analysis process:

Step 1: Establishing the model and network structure
The subject or problem must be clearly and explicitly transformed into a logical system, such as a network. This network structure can be obtained through brainstorming or any other suitable method with the nominal group method. At this stage, the subject is transformed into a network structure in which groups are represented as clusters. Elements within a cluster may be related to one or all elements of other clusters (external dependencies). Elements within a cluster may also be related to each other (internal dependencies).
Step 2: Pairwise comparison matrix and eigenvector
Similar to the binary comparisons performed in AHP, the decision elements in each cluster are compared pairwise based on their importance in relation to the control criteria. The clusters themselves are also compared pairwise based on their role and impact in achieving the goal. Decision-makers must make a decision about binary comparisons of elements or clusters themselves pairwise. In addition, the interdependencies between elements in a cluster must also be compared pairwise. The impact of each element on another element can be represented through the eigenvector. The relative importance of the elements is measured based on a nine-hour quantitative scale.

In this section, the internal importance vector is calculated, which indicates the relative importance of the elements or clusters, and is obtained through the following relationship.

$$\text{AW} = \lambda_{max}\ w_i \tag{12.7}$$

where λmax is the largest eigenvalue of the matrix A. The vector W is obtained using $a = \sum_{i=1}^{n} w_n$, so that the sum of each column in the matrix is equal to one.

To determine the consistency of the comparisons, the criterion weight consistency index is used, which is calculated using the following equation:

$$CI = \frac{\lambda_{max-n}}{n-1} \tag{12.8}$$

In general, if CI is less than 1, the comparison is confirmed.
Step 3: Forming the supermatrix and converting it into a limit supermatrix
To obtain the overall priorities in a system with mutual influences, the internal priority vectors, i.e., the calculated Ws, are entered into the appropriate columns

of a matrix, resulting in a supermatrix, i.e., a partitioned matrix where each part of this supermatrix represents the relationship between two clusters in a matrix. First, the initial supermatrix is constructed.

The standard format of a supermatrix is as follows:

$$
\begin{array}{c} C_1 \qquad C_k \qquad C_n \\ \cdots\; e_1 m_1 \;\cdots\; e_{k1} e_{k2} \;\cdots\; e_{n1} e_{n2} \;\cdots\; e_{n\,mn} \end{array}
$$

$$
w = \begin{array}{cc} C_1 & \begin{array}{c} e_{11} \\ e_{11} \\ e_{13} \\ \vdots \\ e_1 m_1 \end{array} \\ \vdots & \vdots \\ C_k & \begin{array}{c} e_{k1} \\ e_{k2} \\ \vdots \\ e_{knk} \end{array} \\ \vdots & \vdots \\ C_n & \begin{array}{c} e_{n1} \\ e_{n2} \\ \vdots \\ e_{n\,mn} \end{array} \end{array}
\begin{bmatrix} W_{11} & \cdots & W_{1k} & \cdots & W_{1n} \\ \vdots & \vdots & \vdots & \vdots & \vdots \\ W_{k1} & \cdots & W_{kk} & \cdots & W_{kn} \\ \vdots & \vdots & \vdots & \vdots & \vdots \\ W_{n1} & \cdots & W_{nk} & \cdots & W_{nn} \end{bmatrix}
$$

In fact, the columns of the supermatrix are made up of several eigenvectors, the sum of each of which is equal to one. Therefore, it is possible that the sum of each column of the initial supermatrix is greater than one (proportional to the eigenvectors in each column). In order to factor the elements of the column in proportion to their relative weight and make the column sum equal to one, each column of the matrix is standardized. As a result, a new matrix is obtained, and the sum of each of its columns will be equal to one. The new matrix is called the weighted or weighted matrix.

Then, by normalizing the weighted supermatrix, the supermatrix is converted into a random state in terms of columns. In the final step, the limit supermatrix is raised to an odd power by raising all the elements of the weighted supermatrix to an odd power until convergence is achieved. In this case, all the columns of the supermatrix become the same.

$$
\lim_{k \to \infty} W^k \tag{12.9}
$$

Step 4: Calculating the final weight

In the last step, according to the cluster weight and the limit supermatrix, the final weight of the criteria is calculated.

12.3.3 Best-Worst Method (BWM)

In multi-criteria decision-making methods, a number of options are evaluated according to a number of indicators to select the best option. Based on the Best-Worst Method (BWM) presented in 2015, the best and worst indicators are determined by the decision-maker, and a pairwise comparison is made between each of these two criteria (best and worst) and the other indicators. Then, a maximin problem is formulated and solved to determine the weight of the different indicators. Also, in this method, a formula is considered to calculate the inconsistency rate to check the validity of the comparisons. Among the outstanding features of this method compared to other MCDM methods, the following can be mentioned:

The need for less comparative data leads to a more robust comparison, which means that it gives more reliable answers.

Steps of the BWM method:

Step 1: Determine the set of decision-making indicators. In this step, the set of indicators becomes $\{c_1, c_2, \ldots, c_n\}$, which is needed to make a decision.

Step 2: Determine the best (most important, most desirable) and worst (least important and least desirable) indicators. In this step, the decision-maker defines the best and worst indicators in general; no comparison is made at this step.

Step 3: Determine the priority of the best indicator over other indicators with numbers 1 to 9. The vector of the priority of the best indicator over other indicators is shown as follows:

$$A_B = \left(a_{B1}, a_{B2}, \ldots, a_{Bn}\right)$$

In the above vector, a_{Bj} indicates the priority of the best indicator (B) over indicator (j), it is clear that $a_B = 1$.

Step 4: Determine the priority of all indicators over the worst indicator with numbers 1 to 9. The vector of the priority of other indicators over the worst indicator is shown as follows:

$$A_W = \left(a_{1w}, a_{2w}, \ldots, a_{nw}\right)^T$$

In the vector above a_{jw} indicates the preference of index (j) over the worst index (W), it is clear that $a_w = 1$

Step 5: Find the optimal values of the weights $\left(w_1^*, w_2^*, \ldots, w_n^*\right)$. To determine the optimal weight of each of the indices, the pairs $\frac{w_B}{w_j} = a_{Bj}$ and $\frac{w_j}{w_w} = a_{jw}$. To satisfy these conditions at all j, a solution must be found such that the expressions

$\left|\frac{w_B}{w_j}-a_{Bj}\right|$ and $\left|\frac{w_j}{w_w}-a_{jw}\right|$ are effective for all j that are minimized. Considering the non-negativity of the weights and the sum of the weights, the model can be formulated as follows.

$$\min\max\left\{\left|\frac{w_B}{w_j}-a_{Bj}\right|,\left|\frac{w_B}{w_j}-a_{Bj}\right|\right\}$$

$$\text{S.t.}\quad \sum_j w_j = 1 \tag{12.10}$$

$$w_j \geq 0, \quad \textit{for all } j$$

The above model can also be converted to the following model.

$$\min_{s.t.} \xi \tag{12.11}$$

$$\left|\frac{w_j}{w_w}-a_{jw}\right| \leq \xi, \textit{ for all } j$$

$$w_j \geq 0, \quad \textit{for all } j$$

Of course, the linear model of the above function was also presented as follows, and in this study, the weights of the indicators are obtained using the linear model.

$$\min_{s.t.} \xi$$

$$\left|w_B - a_{Bj}w_j\right| \leq \xi, \quad \textit{for all } j$$
$$\left|w_B - a_{Bj}w_j\right| \leq \xi, \quad \textit{for all } j \tag{12.12}$$

$$\sum_j w_j = 1$$

$$w_j \geq 0, \quad \textit{for all } j$$

By solving the above model, the optimal values of ξ^* and $\left(w_1^*, w_2^*, \ldots, w_n^*\right)$ are obtained.

TABLE 12.2
Confirmation of the Security Effects of Electronic Payment Using the Fuzzy Delphi Method

No	Indicator	Decision Score	Status
1	Having sufficient resources	0.736	Approved
2	Employing strong and competent consultants	0.807	Approved
3	Cyberspace management	0.7211	Approved
4	Precise control and monitoring	0.721	Approved
5	Top management support and backing	0.646	Approved
6	Proper management of financial transactions	0.637	Approved
7	Internet server equipment	0.72	Approved
8	Bank	0.741	Approved
9	Website network security	0.637	Approved
10	Data management (accurate, reliable, sufficient, and appropriate information)	0.672	Approved
11	Perceived knowledge	0.763	Approved
12	Perceived security	0.711	Approved

12.4 DATA ANALYSIS

After the research literature and research methodology in the third section, the data analysis is discussed here. The main objective of this research is to analyze the security effects of electronic payment in online shopping using best-worst methods. In this case, the criteria for the security effects of electronic payment in online shopping are discussed, and the weighting of the criteria was examined with the help of the analytic network process technique, and the better criteria were identified with the Best-Worst Method.

To achieve the goal of this research, which is to analyze the security effects of electronic payments in online purchases using best-worst methods, it is necessary to define and identify indicators of the security effects of electronic payments by which the identified factors can be prioritized. Since the number of identified criteria is large, the approved criteria are given in Table 12.2, and we examine them using the fuzzy Delphi method.

The factors were presented in the form of questionnaire and provided to the experts, and the experts responded with their opinions on an optimistic, pessimistic, and probable spectrum on a scale from 1 to 10. In this way, a spectrum of ten is considered for each of the above levels, and the experts expressed their opinions on each indicator. The geometric mean of 15 experts was calculated, and the final score was given to the answers. The accepted number is 0.623.

12.4.1 Analytical Network Process Technique

The ANP method was used for weighting. We formed a correlation matrix and determined the relationship between the criterion and the sub-criteria. To form the

TABLE 12.3
Normalized Criteria Matrix

	Management	Technology-Related	Human Factors	Normalized
Management	1	0.6117	2.132	0.261
Technology-Related	1.621	1	3.109	0.431
Human Factors	0.469	0.322	1	0.14
Inconsistency Rate			= 0.006	

pairwise matrix, the experts' opinions were collected, and the pairwise matrix was formed. A compatibility test was performed for the group pairwise comparison of each matrix, and the weight comparisons and the intensity of the effect of each criterion are calculated in Table 12.3.

After performing the normalized matrix, we form the super matrix. For this purpose, the unweighted super matrix, the matrix, and the weighted super matrix should be formed. The eigenvectors (weights) extracted from the pairwise comparison matrices show the intensity of importance of the criteria. By placing each of the vectors in its appropriate place, the unweighted super matrix is formed. Weight of criteria are Management factors=0.541, Technology-related factors=0.132, and Human factors=0.327.

12.4.2 Best-Worst Technique

After identifying and selecting the criteria and sub-criteria that are effective on the security effects of electronic payments in online purchases, we obtain the weighting of each of the selected criteria and sub-criteria once again using the Best-Worst Method (BWM) and in accordance with the steps mentioned in the previous section and the hierarchical structure.

Therefore, first, the most important and least important criteria and effective sub-criteria should be determined. In a face-to-face meeting with experts, the most important and least important criteria and effective sub-criteria were examined, and finally, among the effective criteria, the criteria of management factors (C1) and technology-related factors (C2) were considered as the most important and least important criteria that are effective on the security effects of electronic payments in online purchases, respectively. Also, among the sub-criteria, the sub-criteria of having sufficient resources (C11) and employing strong and competent consultants (12C) in the managerial factors criterion (C1), the sub-criteria of website network security (23C) and Internet server equipment (C21) in the technology-related factors criterion (C2), and the sub-criteria of perceived security (C32) and perceived knowledge (C31) in the human criterion (3C) were considered as the most important and least important sub-criteria in the security effects criteria of electronic payment in online purchases, respectively.

In the next step, the vector of priority of the most important criteria and sub-criteria (in each criterion) over other criteria and sub-criteria is determined. To determine this vector, experts were asked to indicate the priority of the most important criteria and sub-criteria over other criteria and sub-criteria from 1 to 9. Finally, the average of the collected data was taken, and the results are as follows:

Most important criterion C1 C2 C3
Most important sub-criterion with respect to the first criteria C11 C12 C13 C14 C15 C16
Most important sub-criterion with respect to the second criteria C21 C22 C23 C24
Most important sub-criterion with respect to the third criteria C31 C32

Then the priority vector of other criteria and the sub-criteria are determined relative to the least important criterion and sub-criteria. To determine this vector, the same procedure was followed as in the previous step.

Then the values of the optimal weights are obtained (w_1*, w_2*,..., w_n*). The values of the weights can be obtained as in the following equations:

$$\min_{s.t.} \xi$$

$$\begin{aligned} \left| w_B - a_{Bj} w_j \right| &\leq \xi, \quad \textit{for all } j \\ \left| w_B - a_{Bj} w_j \right| &\leq \xi, \quad \textit{for all } j \end{aligned} \tag{12.13}$$

$$\sum_j w_j = 1$$

$$w_j \geq 0, \quad \textit{for all } j$$

Finally, by solving the above models using LINGO software, for each of the criteria and sub-criteria of the research, the final results are obtained:

ξ* value = 0.115432

Compatibility index = 1.3

Compatibility rate = 0.0887

According to the solution of the linear programming model of the Best-Worst Method, it is observed that among the three criteria, the criteria of management factors (C1), human factors (C3), and system-related factors (C2) are important with weights of 0.682, 0.234, and 0.083, respectively. Also, among the sub-criteria, the sub-criteria of having sufficient resources (C11), perceived security (C32), and support and

TABLE 12.4
Weights Calculated with Two Techniques

Criteria	ANP Technique (a_j)	BWM Technique β_i
Management factors	0.541	0.68
Technology-related factors	0.132	0.09
Human factors	0.327	0.23

patronage of top management (15C) are of the highest importance. Also, the decision-making consistency rate is very good, that is, close to zero.

12.4.3 Calculation of the Final Weight of the Factors

In this part of the research, the final weights of each indicator were calculated using the modified entropy technique, the steps of which are explained below. Table 12.4 shows the weights of the research criteria in each of the network analysis process techniques and the best-worst.

Now, using the following formula, the final combined weights of the three criteria are calculated.

$$final\ w_j = \frac{a_j * \beta_j}{\Sigma a_j * \beta_j}$$

Final weight of criteria:
Management factors = 0.812 rank 1
Technology-related factors = 0.023 rank 3
Human factors = 0.165 rank 2

As can be seen in Figure 12.1, management factors have the highest importance and have a significant difference compared to other criteria, which indicates their excessive importance.

12.5 CONCLUSIONS AND SUGGESTIONS

One of the important parts of the research, which can actually be a way to turn theories into practice for future success, is correct conclusions and relevant and appropriate suggestions. Conclusions based on correct analyses can solve the problems facing the organization for which the research was designed. Today, information security in electronic payments is one of the important issues, and the category of information security should be viewed not as a product but as a process. Undoubtedly, information about threats, attacks, and how to deal with them has a special place in the process of information security, and it is necessary to always keep ourselves updated with the latest information in this field. In a globalizing world, electronic payment is an

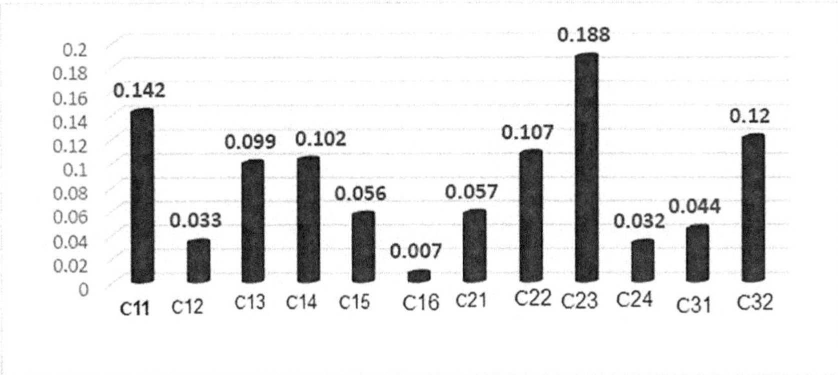

FIGURE 12.1 Weights of sub-criteria.

important aspect of the development of the payment sector. By solving these issues and challenges, the payment sector can make its customers loyal and provide more facilities to them. Payments should eagerly consider the demands of their customers and successfully prioritize the feedback and reactions of the community.

It is obvious that without additional security measures, widespread electronic commerce is not sustainable and viable. However, a competently designed and appropriate electronic payment system can provide better security and flexibility in use than traditional electronic payment systems.

Information security plays a fundamental and decisive role in creating the necessary conditions for the successful implementation of national information and communication technology plans, e-government, e-commerce, and the implementation of projects in the fields of education and training, health, and finance and credit. Today, the security of the digital space portrays a new aspect of the national security of every country. It is hoped that, in parallel with the rapid development of various applications of information and communication technology in various sub-sectors in the program coverage, by recognizing and determining the country's key infrastructures that are vitally dependent on information and then planning, organizing, and investing appropriately to protect these infrastructures, the path of the country's comprehensive development in achieving a knowledge-based society will be paved.

The limitations that the researcher encountered in this research are related to the following three important factors:

- Time constraints
- Participation
- Human resources

Also, because there was no comprehensive view of these systems in the online environment (especially the experts who were willing to cooperate in this research), their understanding, recognition, examination, and evaluation faced several limitations.

It is better to pay attention to the 12 factors raised in this research, because they are very important for analyzing security effects. Accordingly, it is suggested that the organization use more up-to-date and experienced individuals and specialists in this field who can resolve processing errors with greater accuracy and precision as soon as possible. In order to increase the security of its services, it should carry out all support-related processes in the organization with greater care. And the processing error review index is of the greatest importance in the eyes of experts, and the results show that the support index is of the greatest importance among the technology factor criteria in the eyes of experts, and the information rules and processing software index is of the greatest importance in the eyes of experts. Accordingly, it is suggested that, from the experts' point of view, laws were of the greatest importance, so the need to pay attention to developing up-to-date laws and rules in accordance with international standards in this field is evident. The incident cost measurement mechanisms index is of the greatest importance in the eyes of experts. Accordingly, it is suggested that, considering the preferred option, these incidents can play a fundamental role in reducing the efficiency and information security of the organization. Therefore, it seems necessary to develop indicators and criteria that can effectively measure this process. The index of equipment and support facilities deficiencies is of the greatest importance in the opinion of experts. Accordingly, it is suggested that by allocating specific budgets in this field, they try to equip themselves with up-to-date global tools and software. The index of compliance with security implementation standards is of the greatest importance in the opinion of experts. Accordingly, it is suggested that specialized experts in this field be employed to identify the most important global standards. The index of information security events is of the greatest importance in the opinion of experts, and the indicators of security weakness reporting, incident tracking, cost measurement mechanism, and rapid response to incidents are of lesser importance, respectively. Accordingly, it is suggested that, given the important role of information security, codified laws in accordance with global standards should be considered to measure the compliance of measures taken with global security processes.

Suggestions for future research:

- Identifying relationships between criteria using the DEMATEL method.
- The network analysis process method used in this study can be implemented in a fuzzy manner.
- Identifying the factors of this time, the factors of payment security effects, using the factor analysis method.

References

Akbari, F., Fazlollahtabar, H., & Mahdavi, I. (2013). An uncertain decision making process considering customers and services in evaluating banks: A case study. *International Journal of Strategic Decision Sciences (IJSDS)*, *4*(2), 48–78.

Alikhani, M., & Fazlollahtabar, H. (2014). A mathematical model for optimizing organizational learning capability. *Advances in Operations Research*, *2014*(1), 490210.

Arabshahi, H., Fazlollahtabar, H., & Maboudi, L. (2020). Efficiency evaluation of supply chain network using a framework based on DEA and seller-buyer structure. *Asia-Pacific Journal of Operational Research*, *37*(06), 2050029.

Bagheri, E., Fazlollahtabar, H., & Ashoori, M. T. (2018). Product pricing with marketing data under risk using business intelligence. *Journal of Sustainable Competitive Intelligence*, *8*(3), 1–14.

Farjoo, M., Fazlollahtabar, H., & Ashoori, M. T. (2018). Design of a recommender system for online shopping using decision tree and Apriori algorithm. *Journal of Software Engineering & Intelligent Systems*, *3*, 236–244.

Fazlollahtabar, H. (2008). A dynamic programming approach to identifying the shortest path in virtual learning environments. *E-Learning and Digital Media*, *5*(1), 89–96.

Fazlollahtabar, H. (2010). Employing fuzzy logic for a real-time comprehensive quality assessment model of service providers in e-learning environments. In M. Khosropour (Ed.), *Cases on Technology Enhanced Learning through Collaborative Opportunities* (pp. 223–241). IGI Global.

Fazlollahtabar, H. (2012a). Applying information technology in developing business models of export. *International Journal of Business and Systems Research*, *6*(1), 89–107.

Fazlollahtabar, H. (2012b). Reliability-based dynamic programming for e-learning user profile assessment. *International Journal of Information and Communication Technology Education (IJICTE)*, *8*(3), 13–21.

Fazlollahtabar, H. (2012c). Six sigma based integrated mathematical model for optimizing electronic marketing decisions. *International Journal of Customer Relationship Marketing and Management (IJCRMM)*, *3*(2), 61–76.

Fazlollahtabar, H. (2020). A business intelligence and analytics for service oriented manufacturing resource planning. *Information Sciences Letters*, *9*(1), 7–14.

Fazlollahtabar, H. (2021a). An intelligent sales management system based on Internet of Things and Bayesian network. *Industrial Management Perspective, 11*(4), 59–84.

Fazlollahtabar, H. (2021b). Online business quality of service information analytics using artificial intelligence. *Information Sciences Letters*, *10*(1), 93–100.

Fazlollahtabar, H. (2022a). A multidimensional decision making for supplier selection in the presence of information systems. *Croatian Operational Research Review*, *13*(1), 31–47.

Fazlollahtabar, H. (2022b). Intelligent marketing decision model based on customer behavior using integrated possibility theory and K-means algorithm. *Journal of Intelligent Management Decision*, *1*(2), 88–96.

Fazlollahtabar, H. (2024). Comprehensive evaluation of e-business satisfaction variables and indicators in B2C model using the negotiation decision function method. *Journal of Industrial Management Perspective*, *14*(3), 9–23.

Fazlollahtabar, H., Abbasi, A., & Taheri, A. (2010). Designing an intelligent information system for a multilayer electronic supply chain network. *International Journal of Systems Control*, *1*(4), 1–12.

Fazlollahtabar, H., & Aghasi, E. (2012). A two-stage mechanism as a decision support for marketing via customer satisfaction measures. *Journal of the Chinese Institute of Industrial Engineers*, *29*(4), 270–281.

Fazlollahtabar, H., Aghasi, E., & Aghasi, A. (2013). Effect of customer satisfaction measures on a supply chain. *Advances in Industrial Engineering and Management*, *2*, 35–39.

Fazlollahtabar, H., & Ahmadiahangar, R. (2024). Profit analytics in disruption risk for electrical energy supply network considering cost-oriented big data. *Energy & Environment, 36*(7), 3527–3544. 0958305X231225599.

Fazlollahtabar, H., Akbari, F., & Mahdavi, I. (2013). A fuzzy goal programming for optimizing service industry market using virtual intelligent agent. *Journal of Industrial and Production Engineering*, *30*(1), 20–29.

Fazlollahtabar, H., & Ashena, M. (2013). A Bayesian network decision support system for order management in new product development. *Computer Science and Information Technology*, *1*, 82–89.

Fazlollahtabar, H., & Ashoori, M. T. (2020). Business analytics using dynamic pricing based on customer entry-exit rates tradeoff. *Statistics, Optimization & Information Computing*, *8*(1), 272–280.

Fazlollahtabar, H., & Ehsani, M. A. (2010). Integration between regression model and fuzzy logic approach for analyzing various electronic commerce effects on economic growth in organizations. *Journal of Electronic Commerce in Organizations (JECO)*, *8*(4), 17–31.

Fazlollahtabar, H., Hajmohammadi, H., & Mohajeri, A. (2012). Designing an electronic supply chain management system in an electronic market considering customer satisfaction and logistic. *International Journal of Customer Relationship Marketing and Management (IJCRMM)*, *3*(3), 74–88.

Fazlollahtabar, H., & Mahdavi, I. (2012). Optimizing e-shopping sales efficiency using virtual intelligent agent and pricing concept. *Asian Journal of Information and Communications*, *4*(1), 27–35.

Fazlollahtabar, H., & Mohseninejad, M. (2015). A neural network decision support system for analysing markets. *Journal of Advances in Management Sciences & Information Systems*, *1*, 38–46.

Hedayati, A., Shirazi, B., & Fazlollahtabar, H. (2014). An assessment model for the state of organizational readiness in service oriented architecture implementation based on fuzzy logic. *Computer Science and Information Technology*, *2*(1), 1–9.

Khalilinezhad, S., Fazlollahtabar, H., Minaei-Bidgoli, B., & Eslami Nosratabadi, H. (2021). Detecting valuable customers using the trade patterns of financial transactions applying integrated RFM and OLAP. *International Journal of Industrial Engineering and Production Research*, *32*(3), 1–18.

Mahdavi, I., Fazlollahtabar, H., Mahdavi-Amiri, N., & Kakouei, M. (2010). Designing integrated models as a decision aid in web-based marketing. *Journal of Information Technology Applications and Management*, *17*(1), 1–10.

Mahdavi, I., Fazlollahtabar, H., Paydar, M. M., Heidarzade, A., & Hadighi, S. A. (2007). The role of information technology in industry development based on SWOT analysis in Iran. In *Proceedings of the Korea Society of Information Technology Applications Conference* (p. 151). The Korea Society of Information Technology Applications.

Mahdavia, I., Abbaspour-Ghomi, H., Sahebjamnia, N., Eissazadeh-Roshan, Y., Fazlollahtabar, H., Mahdavi-Amiri, N., & Hamzehpour, A. (2009). Strategic development and SWOT analysis for IT/ICT programs in universities of Mazandaran in Iran using digital indices. *Journal of Digital Convergence*, *7*(1), 31–47.

Marzouni, S. B., Fazlollahtabar, H., & Mahdavi, I. (2018). Integrated information exchange for logistics systems based on Internet of Things. *International Journal of Robotics Research*, *3*(011), 1–8.

Mehrmanesh, H., Parikhi, S., & Fazlollahtabar, H. (2013). Modelling the customer relationship management in a multi-layer supply chain considering product life cycle. *International Journal of Services and Operations Management*, *16*(4), 525–547.

Mehrmanesh, H., Pazhoohi, B., & Fazlollahtabar, H. (2013). Proposing a framework for performance improvement in a bi-direction multi-layer and multi-product supply chain using data mining. *Journal of Management Science and Practice*, *1*(1), 22.

Shahri, S., Fazlollahtabar, H., Mahdavi, I., & Rahmani, Z. (2013). Improvement model for strategic planning in the frame of earned value using mathematical clustering. *Universal Journal of Industrial and Business Management, 1*(1), 8–18. Doi: 10.13189/ujibm.2013.010102.

Taheri, M., Shirazi, B., & Fazlollahtabar, H. (2014, December). QoS-based web services composition using water flow-like algorithm. In *Information Engineering* (Vol. 3, pp. 27–38). Science and Engineering Publishing Company.

Taherikhonakdar, Z., & Fazlollahtabar, H. (2024). Integrated Delphi-Interpretive structural modeling for sustainable green software development. *International Journal of Industrial Engineering*, *35*(2), 1–26.

Tajdin, A., Shirazi, B., Sahebjamnia, N., & Fazlollahtabar, H. (2007). Designing an integrated e-learning environment for global education using real-time simulators. In *IADIS International Conference e-Learning 2007 (part of MCCSIS 2007)* (pp. 123–130).

Index

For Product Safety Concerns and Information please contact our EU representative GPSR@taylorandfrancis.com
Taylor & Francis Verlag GmbH, Kaufingerstraße 24, 80331 München, Germany

www.ingramcontent.com/pod-product-compliance
Ingram Content Group UK Ltd.
Pitfield, Milton Keynes, MK11 3LW, UK
UKHW022218190926
475702UK00003B/429

* 9 7 8 1 0 4 1 2 3 7 3 3 4 *